Honda GL1000 Gold Wing Owners Workshop Manual

by Mansur Darlington

With an additional chapter covering 1978-79 models

by Chris Rogers

Models covered

GL 1000, K1 and K2	March 1975 — April 1979 UK only
GL 1000 KZ	April — September 1979 UK only
GL 1000	May 1975 — September 1979 USA only
GL 1000 LTD	1976 USA only

ISBN 978 0 85696 710 8

© J H Haynes & Co. Ltd. 1991

Printed in Malaysia   (309-11Q7)

ABCDE
FGHIJ

2

J H Haynes & Co. Ltd.
Sparkford, Yeovil,
Somerset BA22 7JJ, England

Haynes North America, Inc
859 Lawrence Drive, Newbury Park,
California 91320, USA

Acknowledgements

Our grateful thanks are due to Honda UK Limited for permission to reproduce their line drawings, and to Richard Loder, of Loders Honda Centre, Dorchester, who supplied the machine used in this manual.

Brian Horsfall and Martin Penny gave considerable assistance with the dismantling and rebuilding. Les Brazier arranged and took the photographs, and also assisted in lifting the engine from floor level onto the workbench; no mean task. Illustrations and advice about tyre fitting were kindly supplied by the Avon Rubber Company and NGK Spark plugs (UK) Limited supplied the information about comparative spark plug conditions.

Especially thanks are due to Jeff Clew who gave invaluable advice and edited the text.

About this manual

The author learnt his motorcycle mechanics by trial and error possibly more by error. In presenting this manual it is hoped that these errors can be avoided by others. Only by supervising the work himself, under conditions similar to those in which the amateur mechanic works, can the author ensure that the text is a true and concise record of procedure.

Honda service tools were not used, as generally an alternative method of removing or replacing a part was possible.

These motorcycles are designed to the metric system, and it is far easier to work metric. Metric spanners must be used.

A torque wrench should be begged or borrowed for use where torque wrench settings are given. Some car accessory shops and tool hire companies will often loan one.

Always have all tools and replacement parts to hand before commencing work. Baking trays or similar containers are useful for putting small parts in. Replace nuts and washers on the studs they fitted where possible, this avoids loss. Unless otherwise mentioned, reassembly should be carried out in reverse order to dismantling.

Each of the seven Chapters is divided into numbered sections. Within the sections are numbered paragraphs. Cross-reference throughout this manual is quite straightforward and logical. When reference is made, 'See Section 6.10', it means Section 6, paragraph 10 in the same Chapter. If another Chapter were meant it would say, 'See Chapter 2, Section 6.10'.

All photographs are captioned with a section/paragraph number to which they refer, and are always relevant to the Chapter text adjacent.

Figure numbers (usually line illustrations) appear in numerical order, within a given Chapter. 'Fig. 1.1' therefore refers to the first figure in Chapter 1.

Left-hand and right-hand descriptions of the machine and their components refer to the left and right of a given machine when normally seated.

Motorcycle manufacturers continually make changes to specifications and recommendations, and these, when notified, are incorporated into our manuals at the earliest opportunity.

Whilst every care is taken to ensure that the information in this manual is correct no liability can be accepted by the authors or publishers for loss, damage or injury caused by any errors in or omissions from the information given.

Modifications to the Honda Gold Wing

In the short time that the Gold Wing has been on the market only a small number of minor modifications have been made, none of which materially alter the machines characteristics or the procedure for executing repair work. There are, however, differences between the UK and USA models; mainly in the electrical system and in the type of steering lock fitted. These points are dealt with in the relative Chapters.

Shortly before this publication went to press, Honda announced the introduction of a limited number of 2000 Special Gold Wings for distribution in the USA to mark the Bicentennial anniversary of that nation. These machines are basically the same as the standard model, differing mainly in the care with which the machine is built, the paintwork and the trim. The engines are built from selected components, to give the optimum balance and power.

Contents

Note: General descriptions and specifications are given in each Chapter immediately after the list of contents. Fault diagnosis is given at the end of the Chapter.

1976 999cc GL1000 GOLD WING

Introduction to the Honda Gold Wing

The present Honda empire, which started in a wooden shack in 1947, now occupies a vast modern factory.

The first motorcycle to be imported into the UK in the early 60's, the 250 cc twin 'Dream', was the thin edge of a wedge which has been the Japanese domination of the motorcycle industry. Strange it looked too, to western eyes, with pressed steel frame, and 'square' styling.

In 1959, Honda commenced road racing in Europe, at the IOM TT races. They came 'to learn, next year to race, maybe', but walked off with the manufacturers team award. A few years after this derided start, they were to dominate all classes, with such riders as Mike Hailwood, Jim Redman, and the late Tom Phillis and Bob McIntyre, on four, five and six cylinder machines.

Even the previously unbeaten Italian multis no longer had things their own way, and were hard put to continue racing under really competitive terms.

Honda withdrew finally from racing in 1967, when at the top of the tree. Five years after the introduction of their road going 750 cc Four, Honda brought out the 1000 cc Gold Wing to fill a position in the market not yet catered for by this company. The Gold Wing was designed as the ultimate in long distance touring machines, capable of carrying two people comfortably over unlimited distances, with a large load of luggage. As will be seen from the overall specification, the Gold Wing has a standard of sophistication that makes it unique.

Model dimensions and data

Overall length	90.8 in. (2305 mm)
Overall width	34.4 in. (875 mm)
Height	48.2 in. (1225 mm)
Wheelbase	60.9 in. (1545 mm)
Ground clearance	5.9 in. (81 mm)
Weight (dry)	584 lb (265 kg)
Engine weight	234 lb (106.1 kg)

Ordering spare parts

When ordering spare parts for any Honda, it is advisable to deal direct with an official Honda agent, who should be able to supply most items ex-stock. Parts cannot be obtained from Honda (UK) Limited direct; all orders must be routed via an approved agent, even if the parts required are not held in stock.

Always quote the engine and frame numbers in full, and colour when painted parts are required.

The frame number is located on the side of the steering head, and the engine number is stamped on the crankcase immediately to the rear of the oil pressure warning switch.

Use only parts of genuine Honda manufacture. Pattern parts

are available, some of which originate from Japan, but in many instances they may have an adverse effect on performance and/or reliability.

Honda do not operate a 'service exchange' scheme.

Some of the more expendable parts such as spark plugs, bulbs, tyres, oils and greases etc., can be obtained from accessory shops and motor factors, who have convenient opening hours, charge lower prices and can often be found not far from home. It is also possible to obtain parts on a Mail Order basis from a number of specialists who advertise regularly in the motorcycle magazines.

Frame number location

Engine number location

Safety first!

Professional motor mechanics are trained in safe working procedures. However enthusiastic you may be about getting on with the job in hand, do take the time to ensure that your safety is not put at risk. A moment's lack of attention can result in an accident, as can failure to observe certain elementary precautions.

There will always be new ways of having accidents, and the following points do not pretend to be a comprehensive list of all dangers; they are intended rather to make you aware of the risks and to encourage a safety-conscious approach to all work you carry out on your vehicle.

Essential DOs and DON'Ts

DON'T start the engine without first ascertaining that the transmission is in neutral.

DON'T suddenly remove the filler cap from a hot cooling system – cover it with a cloth and release the pressure gradually first, or you may get scalded by escaping coolant.

DON'T attempt to drain oil until you are sure it has cooled sufficiently to avoid scalding you.

DON'T grasp any part of the engine, exhaust or silencer without first ascertaining that it is sufficiently cool to avoid burning you.

DON'T allow brake fluid or antifreeze to contact the machine's paintwork or plastic components.

DON'T syphon toxic liquids such as fuel, brake fluid or antifreeze by mouth, or allow them to remain on your skin.

DON'T inhale dust – it may be injurious to health (see *Asbestos* heading).

DON'T allow any spilt oil or grease to remain on the floor – wipe it up straight away, before someone slips on it.

DON'T use ill-fitting spanners or other tools which may slip and cause injury.

DON'T attempt to lift a heavy component which may be beyond your capability – get assistance.

DON'T rush to finish a job, or take unverified short cuts.

DON'T allow children or animals in or around an unattended vehicle.

DON'T inflate a tyre to a pressure above the recommended maximum. Apart from overstressing the carcase and wheel rim, in extreme cases the tyre may blow off forcibly.

DO ensure that the machine is supported securely at all times. This is especially important when the machine is blocked up to aid wheel or fork removal.

DO take care when attempting to slacken a stubborn nut or bolt. It is generally better to pull on a spanner, rather than push, so that if slippage occurs you fall away from the machine rather than on to it.

DO wear eye protection when using power tools such as drill, sander, bench grinder etc.

DO use a barrier cream on your hands prior to undertaking dirty jobs – it will protect your skin from infection as well as making the dirt easier to remove afterwards; but make sure your hands aren't left slippery. Note that long-term contact with used engine oil can be a health hazard.

DO keep loose clothing (cuffs, tie etc) and long hair well out of the way of moving mechanical parts.

DO remove rings, wristwatch etc, before working on the vehicle – especially the electrical system.

DO keep your work area tidy – it is only too easy to fall over articles left lying around.

DO exercise caution when compressing springs for removal or installation. Ensure that the tension is applied and released in a controlled manner, using suitable tools which preclude the possibility of the spring escaping violently.

DO ensure that any lifting tackle used has a safe working load rating adequate for the job.

DO get someone to check periodically that all is well, when working alone on the vehicle.

DO carry out work in a logical sequence and check that everything is correctly assembled and tightened afterwards.

DO remember that your vehicle's safety affects that of yourself and others. If in doubt on any point, get specialist advice.

IF, in spite of following these precautions, you are unfortunate enough to injure yourself, seek medical attention as soon as possible.

Asbestos

Certain friction, insulating, sealing, and other products – such as brake linings, clutch linings, gaskets, etc – contain asbestos. *Extreme care must be taken to avoid inhalation of dust from such products since it is hazardous to health.* If in doubt, assume that they *do* contain asbestos.

Fire

Remember at all times that petrol (gasoline) is highly flammable. Never smoke, or have any kind of naked flame around, when working on the vehicle. But the risk does not end there – a spark caused by an electrical short-circuit, by two metal surfaces contacting each other, by careless use of tools, or even by static electricity built up in your body under certain conditions, can ignite petrol vapour, which in a confined space is highly explosive.

Always disconnect the battery earth (ground) terminal before working on any part of the fuel or electrical system, and never risk spilling fuel on to a hot engine or exhaust.

It is recommended that a fire extinguisher of a type suitable for fuel and electrical fires is kept handy in the garage or workplace at all times. Never try to extinguish a fuel or electrical fire with water.

Note: *Any reference to a 'torch' appearing in this manual should always be taken to mean a hand-held battery-operated electric lamp or flashlight. It does **not** mean a welding/gas torch or blowlamp.*

Fumes

Certain fumes are highly toxic and can quickly cause unconsciousness and even death if inhaled to any extent. Petrol (gasoline) vapour comes into this category, as do the vapours from certain solvents such as trichloroethylene. Any draining or pouring of such volatile fluids should be done in a well ventilated area.

When using cleaning fluids and solvents, read the instructions carefully. Never use materials from unmarked containers – they may give off poisonous vapours.

Never run the engine of a motor vehicle in an enclosed space such as a garage. Exhaust fumes contain carbon monoxide which is extremely poisonous; if you need to run the engine, always do so in the open air or at least have the rear of the vehicle outside the workplace.

The battery

Never cause a spark, or allow a naked light, near the vehicle's battery. It will normally be giving off a certain amount of hydrogen gas, which is highly explosive.

Always disconnect the battery earth (ground) terminal before working on the fuel or electrical systems.

If possible, loosen the filler plugs or cover when charging the battery from an external source. Do not charge at an excessive rate or the battery may burst.

Take care when topping up and when carrying the battery. The acid electrolyte, even when diluted, is very corrosive and should not be allowed to contact the eyes or skin.

If you ever need to prepare electrolyte yourself, always add the acid slowly to the water, and never the other way round. Protect against splashes by wearing rubber gloves and goggles.

Mains electricity and electrical equipment

When using an electric power tool, inspection light etc, always ensure that the appliance is correctly connected to its plug and that, where necessary, it is properly earthed (grounded). Do not use such appliances in damp conditions and, again, beware of creating a spark or applying excessive heat in the vicinity of fuel or fuel vapour. Also ensure that the appliances meet the relevant national safety standards.

Ignition HT voltage

A severe electric shock can result from touching certain parts of the ignition system, such as the HT leads, when the engine is running or being cranked, particularly if components are damp or the insulation is defective. Where an electronic ignition system is fitted, the HT voltage is much higher and could prove fatal.

Routine maintenance

Periodic routine maintenance is essential to keep the motor cycle in a peak and safe condition. Routine maintenance also saves money because it provides the opportunity to detect and remedy a fault before it develops further and does more damage. Maintenance should be undertaken on either a calendar or mileage basis depending on which ever comes sooner. The period between maintenance tasks serves only as a guide since there are many variables eg; age of machine, riding technique and adverse conditions.

The maintenance instructions are generally those recommended by the manufacturer but are supplemented by additional tasks which, through practical experience, the author recommends should be carried out at the intervals suggested. The additional tasks are indicated by an asterisk and are primarily of a preventative nature, which will assist in eliminating unexpected failure of a component or system, due to fair wear and tear, and increase safety margins when riding.

All the maintenance tasks are described in detail together with the procedures required for accomplishing them. If necessary, more general information on each topic can be found in the relevant Chapter within the main text.

Where possible, the use of service tools has been avoided in the Routine maintenance procedures. There are, however, certain operations where the use of service tools is imperative, either because an operation cannot be carried out without the necessary tool, or where there is a risk of damage to a component. It is advised that apart from a good selection of hand tools, which should include a range of metric ring spanners or combination spanners and large and small cross-head screwdrivers, a number of more specialised hand tools be obtained. A 'power' or impact screwdriver is absolutely invaluable for releasing the various casing screws, particularly if the engine has not been dismantled or overhauled since leaving the factory. It is no exaggeration to say that some screws are impossible to remove safely without the aid of this tool. Two pairs of circlip pliers should also be acquired. One internal opening, the other external opening, with 90° jaws. Although hand tools are relatively expensive, the damage inflicted by the use of incorrect or ill-fitting tools can quickly exceed the initial outlay in terms of replacement parts required to make good the damage. Bear in mind also that an enviable collection of tools can be bought with even a small part of the money saved on labour by carrying out the work at home!

1 Tyres and wheels

Check the tyre pressures. Always check the pressure when the tyres are cold as the heat generated when the machine has been ridden can increase the pressures by as much as 8 psi, giving a totally inaccurate reading. Variations in pressure of as little as 2 psi may alter certain handling characteristics. It is therefore recommended that whatever type of pressure gauge is used, should be checked occasionally to ensure accurate readings. Do not put absolute faith in 'free air' gauges at garages or petrol stations. They have been known to be in error.

Inspect the tyre treads for cracking or evidence that the outer rubber is leaving the inner cover. Also check the tyre walls for splitting or perishing. Carefully inspect the treads for stones, flints or shrapnel which may have become imbedded and be slowly working their way towards the inner tube. Remove such objects with a suitable tool. The thing for getting stones out of horses hooves is ideal!

Check each spoke individually for tension by tapping with a metal object. Any marked difference in the pitch of the noise generated indicates a loose spoke. Bear in mind that a spoke that requires excessive tightening will protrude through the wheel rim and may penetrate the inner tube. In this case the tyre will have to be removed and the spoke end filed flush with the nipple head.

2 Battery

Check the electrolyte level in the battery and replenish, if necessary, with distilled water. Do not use tap water as this will reduce the life of the battery. If the battery is removed for filling, note the tracking of the battery breather pipe, which should be replaced in the same position, ensuring that the pipe is not kinked or blocked. If the breather pipe is restricted and the battery overheats for any reason, the pressure produced may, in extreme cases, cause the battery case to fail and a liberal amount of sulphuric acid to be deposited on the electrical harness and frame parts.

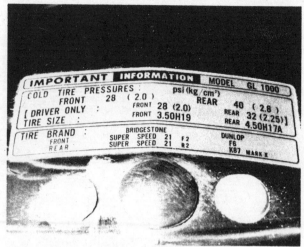

Always maintain tyre pressures correctly

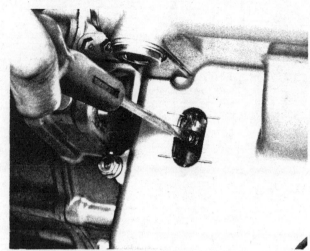

Clean sight glass to view oil level and ...

Check electrolyte level in battery

... top up engine, if necessary

3 Engine oil

Check the engine oil level by viewing through the sight glass in the right-hand crankcase. If necessary, rotate the 'wiper' pad by means of a screwdriver applied to the screw in the centre of the window. This will remove any carbon build-up which may obscure the oil level. Replenish the engine oil so that the level is up to the upper level mark. The correct oil grade is SAE 10W-40.

4 Cooling system

Open the right-hand tank cover. With the engine running and at normal running temperature check the coolant level in the radiator reserve tank. If necessary, remove the reservoir cap and replenish the coolant up to the 'Full' level line. The coolant comprises water mixed with an ethylene glycol base anti-freeze, in a 50/50 ratio. Ideally, distilled water should be used in the cooling system, to reduce corrosion and 'furring-up'. Tap water can be used, particularly if it is known to be soft. For ease of maintenance, a supply of coolant mixture can be made up and used to replenish the system, as necessary. Do not allow the anti-freeze constituent to fall below 40% as the anti-corrosion properties of the coolant mixture will be reduced below an acceptable level. Ensure also that the anti-freeze used is of a type acceptable for use with an aluminium engine. **Warning:** Do not remove the radiator cap when the engine is hot as the reduction in pressure will cause the coolant to boil and exit through the filler orifice.

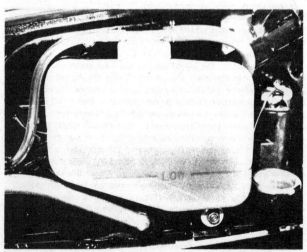

Maintain coolant level in reservoir

5 Brake fluid

Check the hydraulic fluid level of both the front brake master cylinder and the rear brake master cylinder reservoirs. The fluid should lie between the upper and lower lines on the reservoir body. Replenish, if necessary, with hydraulic brake fluid of the correct specification, which is DOT 3 (USA) or SAEJ1703. If the level of fluid in either of the reservoirs is excessively low, check the pads for wear. If the pads are not worn, suspect a fluid leakage in the system. This must be rectified immediately.

6 Safety inspection

Give the whole machine a close visual inspection, checking for loose nuts and fittings, frayed control cables and damaged brake hoses etc.

7 Legal inspection

Check that the lights, horn and flashing indicators function correctly, also the speedometer.

Six monthly or every 6,000 miles

Carry out the tasks listed in the 200 mile/weekly maintenance Section and then complete the following:

1 Engine lubrication

The engine oil requires renewal after 6,000 miles. Drain the old oil after the engine has been running, or after bringing the engine up to normal running temperature. This will thin the oil and improve the draining rate. Place a container of more than 3.2/2.6 US/Imp. quart (3.0 litre) capacity below the front of the engine. Remove the oil filler cap followed by the oil drain plug, which is located below the filter housing. Allow the oil to drain completely and then remove the oil filter housing, complete with element.

Remove the oil filter element and thoroughly clean the inside of the filter housing. Refit a new filter and replace the housing, followed by the drain plug. Note that the filter housing must be replaced in a special position, this being when the aligning marks on the housing align with similar marks on the crankcase - adjacent to the radiator bottom hose manifold.

Replenish the engine through the filler orifice with approximately 3.0 litres of SAE 10W-40 motor oil until the oil level reaches the upper mark in the sight glass. The machine must be standing on level ground when this check is made.

2 Cooling system

Inspect the radiator core externally for clogging by leaves, flies or road dirt. If required, remove obstructions using a high pressure air hose. Bent fins can be straightened carefully, using a suitable tool such as a screwdriver.

The cooling system is sealed and pressurised and as such if rapid reduction of the coolant level is observed, leakage should be suspected. Check the hoses for cracking or splitting, particularly where the screw clips exert pressure. Check also for leakage at the thermostat feed pipes and the coolant drain plug. If leakage is discovered, it will be necessary to drain the cooling system as follows, before repair can take place. Seepage around a hose at a union may be cured by slight tightening at the relevant hose clip.

Place a **clean** container of more than 3.4/2.8 US/Imp. quart (3.2 litre) capacity below the water pump housing at the front of the engine. Remove the radiator pressure cap by pressing the cap downwards and rotating in an anti-clockwise direction. Do not remove the radiator cap when the engine is still hot. Remove the drain plug from the water pump cover and allow the coolant to drain. The cause of leakage can now be remedied. If seepage has occurred at the thermostat feed pipes, the 'O' rings should be renewed. Refill the radiator slowly, to allow as much of the air in the system to escape as possible. Allow the engine to run with the radiator cap off for a few moments until all the air is bled from the system (the level will drop). Finally, refill up to the bottom of the filler neck and replace the pressure cap.

Check fluid level in BOTH master cylinder reservoirs

Fit new filter element replacing ...

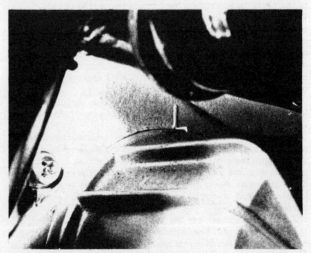

... the housing so that the marks line up

3 Air filter

Open all three of the dummy tank sections and lift the tool tray from position. Slacken the air filter cover wing nut and remove the cover. Lift the air filter element from position.

Tap the element gently to remove any loose dust and then use an air hose to remove the remainder of the dust. Apply the air current from the inside of the element only. If an air hose is not available a tyre pump can be utilised instead. If the corrugated paper element is damp, oily or beginning to disintegrate, it must be renewed. Do not run the engine with the element removed as the weak mixture caused may result in engine overheating and damage to the cylinders and pistons. A weak mixture can also result if the rubber sealing rings on the element are perished or omitted. When replacing the filter assembly, note that the pressed steel cover can only be fitted with the arrow pointing forwards.

4 Contact breaker: examination and adjustment

Remove the contact breaker assembly cover from the rear of the left-hand cylinder head. Remove also all four spark plugs and the small screw cover from the alternator casing.

Examine the points faces of the two sets of contact breaker units. Provided that the points are in good condition they can be cleaned in-situ, using fine grade emery paper backed by a narrow strip of tin, followed by a clean cloth to which methylated spirits or carbon tetrachloride has been applied. The latter chemical should only be used in a well ventilated place.

The points should now be adjusted to restore the correct gap as follows: Using a spanner applied to the alternator rotor centre bolt, rotate the engine in the normal direction of running (turn the alternator bolt clockwise), until the left-hand contact breaker points are fully open. Note that the cam has two lobes, one of which is punch marked. Use only the punch marked cam lobe when adjusting the points gap. Insert a feeler gauge between the points and check that the gap is within the range 0.012 - 0.016 in (0.3 - 0.4 mm). If the gap is incorrect, slacken the two fixed point holding screws and move the point closer to or further away from the moving point until the gap is 0.012 in (0.3 mm). Tighten the two screws and recheck. It is important that the points are in the fully open position or a false reading will be made. Rotate the engine further until the right-hand contact breaker points are in the fully open position. Check and reset as for the left-hand unit.

If the points are badly pitted or burnt they should be removed and dressed, as described in Chapter 4, Section 5.

5 Ignition timing

The ignition timing should be checked and reset only after contact breaker adjustment has been carried out as above. Ignition timing can be carried out manually as described below, or, if a stroboscope is available, when the engine is running as described in Chapter 4, Section 8. Setting ignition timing manually will give quite acceptably accurate results, provided care is exercised. The only real advantage of a stroboscope is that the operation can be accomplished more quickly and the timing can be checked throughout the rev. range.

Remove the flywheel timing mark cover plug, which is located on the upper rear portion of the crankcase, below the fuel filter.

Rotate the engine, again using a spanner on the alternator rotor centre bolt, until the 'F-1' mark on the flywheel is precisely aligned with the index marks on the viewing orifice. It may be easier to align the mark with a thin piece of stiff wire placed across the index marks on the orifice. Cut the wire to a suitable length so that it cannot be easily displaced and so fall into the crankcase.

When the flywheel is in the correct position, as described, the left-hand contact breaker points should just be on the verge of opening. This set of points feeds cylinders No. 1 and No. 2. The exact point at which the contact breakers open can be easily ascertained if a twelve volt bulb is connected between the moving contact point and an earthing position on the engine. With the ignition switched on the bulb will illuminate when the points

Air filter housed in box within dummy fuel tank

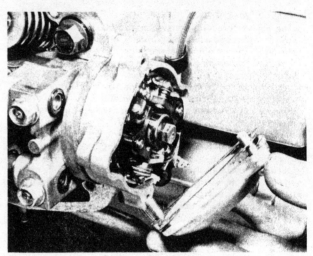
Remove contact breaker assembly cover to ...

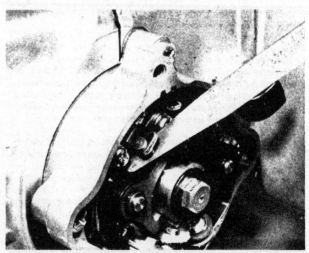
... allow inspection and adjustment of points

open. If the timing is incorrect, slacken off the two screws which clamp the main stator plate in place. Rotate the plate clockwise or anti-clockwise until the timing bulb flickers. At this point the ignition timing is correct on cylinders No. 1 and No. 2. Tighten the stator plate screws and recheck. Rotate the engine through 180° (½ turn) until the 'F-2' timing mark on the flywheel is correctly aligned with the index mark. Check the ignition timing on the right-hand contact breaker unit (cylinders No. 3 and No. 4) as described previously. Note that adjustment of the ignition timing on the right-hand points is made by slackening off the two screws which retain the large half-plate upon which the contact breaker unit rests.

Before replacing the contact breaker assembly cover, apply a few drops of light oil to the felt wick which lubricates the cam. Replace the flywheel timing mark plug before starting the engine.

6 Valve clearance adjustment

Check the valve clearance on both cylinder heads and adjust, if necessary, as follows:

Remove the cylinder head cover from each cylinder head and unscrew the spark plugs.

Rotate the engine until both valves are in the closed position on No. 1 cylinder and the 'T-1' mark on the flywheel is aligned with the index mark. With the crankshaft in this position the No. 1 piston is at TDC. Check the following valves by placing a 0.004 in (0.1 mm) feeler gauge between the valve stem head and the adjuster screw.

No. 1 inlet and exhaust
No. 3 exhaust
No. 4 inlet

If the gap on any valve is incorrect, loosen the locknut on the adjuster and screw the adjuster in or out as necessary. When the gap is correct, prevent the screw rotating by using a screwdriver and tighten the locknut. Rotate the engine through 360° until the T-1 mark reappears and is aligned with the index marks. No. 2 cylinder is now at TDC on the compression stroke. Check and adjust, where necessary, the valve clearances on the following valves.

No. 2 inlet and exhaust
No. 3 inlet
No. 4 exhaust

Check that all the adjuster screw locknuts are tightened fully to 9 - 12 lbs ft (120 - 160 kg cm) before replacing the cylinder head covers. These adjustments must be made with the engine **cold.**

7 Spark plugs

Remove, clean and adjust the spark plugs. Carbon and other deposits can be removed, using a wire brush, and emery paper or a file used to clean the electrodes prior to adjusting the gaps. Probably the best method of spark plug cleaning is by having them shot blasted in a special machine. This type of machine is used by most garages. If the outer electrode of a plug is excessively worn (indicated by a step in the underside) the plug should be renewed. Adjust the points gap on each plug by bending the outer electrode only, so that the gap is within the range 0.024 - 0.028 in (0.6 - 0.7 mm). Before replacing the plugs, smear the threads with graphited grease; this will aid subsequent removal. If replacement plugs are required the correct types are NGK D - 8ESL or Nippon Denso X24ES.

8 Carburettor adjustment

In order to maintain optimum engine performance and fuel economy the carburettors must always be correctly adjusted, and synchronised in relation to each other. Owing to the sophistication of the carburettors the only satisfactory method of adjustment is by means of a set of vacuum dial gauges which are connected to the inlet manifold of each cylinder and are used to measure degrees of negative pressure within each manifold. It is unlikely that the average owner will have access to such instruments or the proficiency with which to carry out the necessary operations. It is strongly advised that the machine be returned to a Honda service agent for carburettor adjustment and synchronisation, whenever necessary.

If a set of vacuum gauges is available the correct degree of vacuum at 1000 rpm is within the range 21 ± 2.5 cm Hg.

9 Throttle adjustment

Adjust the throttle cable so that there is 10° - 15° free play as measured on the periphery of the throttle twist grip, before the throttle begins to open. Adjustment is made by rotating the adjuster on the cable at the twist grip end. Note the locknut, which must be tightened after adjustment.

10 Fuel lines

Check the fuel lines from the petrol tap to the fuel pump and to the carburettors. If leakage is apparent where the reinforced lines connect at the unions, the hose clips may be tightened. Inspect all lines for chaffing or perishing and renew, if suspect.

11 Final drive assembly

Place the machine on the centre stand and remove the oil filler cap from the final drive gear housing. The oil level should be up to the filler orifice. Replenish, if necessary, with a hypoid gear oil conforming to GL-5 specification. The viscosity of the oil is fairly critical and should be as follows for different atmospheric temperatures: Above 5°C (41°F) SAE 90
Below 5°C (41°F) SAE 80

Prise the rubber boot off the drive shaft torque tube and check the universal joint for wear. This can be done by holding the gearbox end of the joint and rotating the rear wheel backwards and forwards. There should be no movement between the two 'knuckles' of the joint. If wear is evident, the drive shaft should be renewed. The splined portions of the drive shaft and final drive pinion shaft should also be checked for wear and lubricated, if necessary. This operation and also renewal of the drive shaft, if necessary, due to wear, requires a considerable amount of dismantling. See Chapter 6, Section 15 for the procedures for accomplishing these operations.

12 Main and side stand

Examine the main (centre) stand and side stand for cracks or bending and lubricate the pivots with a multi-purpose or graphited grease. The centre stand pivots on a hollow shaft, retained at one end by a split pin. The side stand pivots about a single shouldered bolt. Check the return springs and renew them, if weak or strained.

Inspect the rubber pad on the side stand for wear. If it is worn down to or past the wear mark, it should be renewed. Renew with a pad marked '260 lbs'.

Bent or cracked stands can usually be repaired by heating or welding. It is important that the stands on a machine of this weight are in good order. This is particularly so as the cylinder heads are considerably outboard of the frame and consequently vulnerable, if the machine is dropped.

13 Front and rear suspension

Check the suspension systems for correct operation and wear. If there has been any evidence of malfunctioning of the suspension during normal use, the relevant components should be dismantled and overhauled as described in Chapter 6.

Every 12 months or 12,000 miles

Carry out the tasks listed under the weekly/200 mile and 6 months/6,000 headings except where they clash with those listed below, and then complete the following:

1 Spark plugs

Renew the spark plugs as by this mileage efficiency and

reliability will be reduced. The correct plug grade is NGK D-8ESL or Nippon Denso X24ES.

Set the gaps at 0.024 - 0.028 in (0.6 - 0.7 mm) before smearing the threads with graphite grease and replacing the plugs.

2 Air filter

Open all three sections of the dummy fuel tank and remove the tool tray. Remove the air filter cover which is held by a wing nut. Lift out the air filter element and replace it by a new component. The correct dry corrugated paper element is No. 17211-371-003.

3 Steering head bearings

Place the machine on the centre stand so that the front wheel is clear of the ground. This may necessitate placing blocks under the centre stand. Check the adjustment of the steering head bearings by grasping the forks near the front wheel spindle and pulling and pushing them horizontally in a fore and aft direction. Any movement felt between the steering head lug and the fork yokes indicate that the steering head bearings require adjustment, as follows.

Slacken the pinch bolt which passes through the rear of the upper (crown) yoke. Using a suitable 'C' spanner rotate the slotted adjuster ring in a clockwise direction. The adjuster ring lies on the top of the upper yoke, just below the centre of the handlebars. Turn the adjuster ring until all play has just been taken up. Do not overtighten the head races as this will produce an unpleasant rolling effect at low speeds and in some cases speed wobble. If the adjustment is correct, the forks should move onto full lock from the central position when the handlebar is tapped at either end. If, when the adjustment is correct, the bearings feel rough or notchy when the handlebars are moved, it indicates that the ball races have become worn or pitted or in extreme cases have fractured. In either event the steering head bearings should be renewed, necessitating complete removal of the forks as described in Chapter 5, Section 2.

Every 24 months or 24,000 miles

Complete all the operations listed under the previous three Routine maintenance headings and then complete the following:

1 Fuel filter

Remove the fuel filter, which is fitted to the fuel line between the petrol tap and fuel pump. The hoses are retained by screw clips. The filter is a sealed unit and must be renewed.

2 Brake fluid

Honda recommend that the brake fluid in both front and rear brake systems be renewed at this service interval. It is practical to combine this service operation with a complete internal and external examination of the brake operating components, overhauling and renewing where necessary. Refer to Chapter 6 for details of these operations.

3 Front fork clamping oil

Remove the drain plug from the lower leg of each fork leg assembly and allow all the oil to drain. Work the forks up and down to aid draining. Avoid spilling oil onto the tyre rubber or brake discs. Place wooden blocks or a jack below the engine to take the weight of the machine when the fork caps are removed. Prise the rubber covers from the top of each fork leg and using a suitable socket wrench remove the top caps. Replace and tighten the drain plugs and refill each fork leg with 170 - 180 cc (5.8 - 6.1 fl oz) of Automatic Transmission Fluid (ATF) or fork oil. Replace the top caps and rubber covers.

4 Final drive lubrication

Place a suitable container below the final drive gear housing and remove the filler plug, followed by the drain plug. Allow all the oil to drain and replace the drain plug. Refill the gear housing with 200 - 220 cc (6.8 - 7.5 fl oz) of hypoid gear oil conforming to GL-5 specification. If the normal atmospheric temperature under which the machine operates is less than 5ºC (41ºF), use an SAE 80 oil. If the normal temperature is greater than this, use SAE 90 oil. Take great care when refilling not to allow foreign matter into the housing and avoid spilling the lubricant on the rear tyre or brake components.

Additional routine maintenance

1 Brake pads: examination and replacement

The rate of brake pad wear is dependent on the conditions under which the machine operates, weight carried and the style of riding, consequently it is difficult to advise on specific inspection intervals. Whatever inspection interval is chosen, bear in mind that the rate of wear will not be constant.

To check wear on the front brake pads examine the pads through the small window in the main caliper units. If the red mark on the periphery of any pad has been reached, both pads in that set must be renewed. The rate of wear of the two sets are similar so it is probable that they will require renewal at the same time in any case.

Check the rear brake pad wear after removing the plastic caliper cover from position. If the red tongues on the pads have closed together sufficiently that they are within the area marked red on the caliper, they must be renewed.

2 Front brake pad renewal

Remove each brake set individually, using an identical procedure as follows:

Unscrew and remove the two large socket screws which clamp the caliper unit together. Pull the outer and inner portions of the caliper unit from position. The outer portion is still attached to the brake hose. Lift the old pads out. Install the new pads and also the shim which fits against the outer face of the outer pad. The shim must be fitted so that the arrow is in the forward-most position, pointing in an upward direction. Refit the caliper halves and replace the socket screws. It may be necessary to push the caliper cylinder piston inwards to give the necessary clearance. If required, the bleed screw on the caliper can be slackened at the same time as the piston is pushed inwards. This will allow a small amount of fluid to seep out and the piston to move. Place a rag around the bleed screw to prevent the fluid leaking onto the caliper unit. Operate the brake lever, after pad replacement, to check free movement of the pads and to allow the pads to self-adjust.

3 Rear brake pad replacement

Remove the caliper access cover, after unscrewing the single retaining screw. Depress the flat spring which locates both pad pins and remove the upper pin. Tilt the spring back and pull out the lower pin. The brake pad assembly can now be lifted from position.

Install a set of new pads, together with the two shims, the arrows on which must face downwards. Replace the pad pins, ensuring that the pin locating spring locates correctly in the 'waisted' sections of the two pins. Refit the caliper cover and apply the brake to adjust automatically.

4 Clutch adjustment

In common with brake pad wear, clutch wear and the resultant necessary adjustment depends on operating conditions and the style of riding. Adjust the clutch, when necessary, as follows: Remove the clutch adjustment cover which is retained by two bolts and is located at the rear of the engine.

Loosen the locknuts on the upper cable adjuster screw and lower adjuster screw and screw both adjusters in as far as possible. Loosen the locknut on the end of the clutch operating shaft and using a screwdriver, turn the shaft in a clockwise direction until it becomes hard to turn. Rotate the shaft back about ¾ turn and tighten the locknut. Replace the clutch cover.

Unscrew the cable lower adjuster until play of 0.2 - 0.6 in (5 - 15 mm) can be felt, measured at the ball end of the handlebar lever. Tighten the locknut. Fine adjustment can be made using the upper adjustment.

Adjust clutch after removing inspection cover

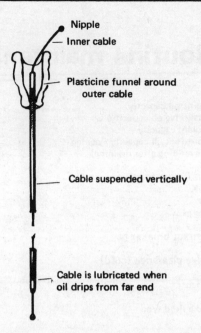

Nipple

Inner cable

Plasticine funnel around outer cable

Cable suspended vertically

Cable is lubricated when oil drips from far end

Oiling a control cable

Routine maintenance capacities and data

Engine oil capacity	3.2/2.6 US/Imp. quart (3.0 litre)
Final drive oil capacity	200 - 220 cc (6.8 - 7.5 fl oz)
Coolant capacity	3.4/2.8 US/Imp. quart (3.2 litre)
Front fork oil capacity - per leg	170 - 180 cc
(for refilling after draining)	

Spark plug type
NGK	D - 8ESL
ND	X24ES
Plug gap	0.024 - 0.026 in (0.6 - 0.7 mm)

Contact breaker gap
Contact breaker gap	0.012 - 0.016 in (0.3 - 0.4 mm)

Valve clearance (cold)
Inlet	0.004 in (0.1 mm)
Exhaust	0.004 in (0.1 mm)

Brake fluid type	DOT 3 (USA)
	SAE J1703 (UK)

Tyre pressures (solo)	Front 28 psi (2.0 kg/cm^2)
	Rear 32 psi (2.25 kg/cm^2)
Tyre pressures (with passenger)	Front 28 psi (2.0 kg/cm^2)
	Rear 40 psi (2.8 kg/cm^2)

Recommended lubricants

Component	Grade or type of lubricant	Quantity
Engine	SAE 10W/40 or 10W/50 oil	3.1 Imp quarts, 3.7 US quarts or 3.5 litres
Final drive	Hypoid gear oil	210 cc
Front fork oil (per leg)	Automatic transmission fluid	175 cc (refill) 200 cc (dry)
Brake fluid	DOT 3	
Engine assembly	Molybdenum disulphide grease	
Oil seals and 'O' rings	Silicon spray	
Wheel bearings, steering head bearings etc.	High melting point grease	

Chapter 1 Engine, clutch and gearbox

Contents

Specifications

Engine

Type	Horizontally-opposed, water cooled, four cylinder
Bore	72 mm (2.834 in)
Stroke	61.4 mm (2.417 in)
Comp. ratio	9.2 : 1
Capacity	999 cc (61.0 cu. in)
Bhp	80 @ 7000 rpm
Engine rotation	Clockwise (viewed from front of engine)

Valve operation	Single overhead camshaft, toothed belt drive
Valve timing:	
Inlet opens	5° BTDC
Inlet closes	50° ABDC
Exhaust opens	50° BBDC
Exhaust closes	5° ATDC
Valve clearance (cold):	
Inlet	0.004 in (0.1 mm)
Exhaust	0.004 in (0.1 mm)
Cylinder head:	
Maximum warpage	0.0039 in (0.1 mm)

Pistons and rings

Piston skirt outside diameter	2.8325 - 2.8335 in (71.945 - 71.97 mm)
Wear limit	2.8288 in (71.85 mm)
Ring groove width:	
Top and second	0.0591 - 0.0599 in (1.5 - 1.52 mm)
Wear limit	0.0630 in (1.6 mm)
Oil ring	0.1104 - 0.1110 in (2.805 - 2.82 mm)
Wear limit	0.1142 in (2.9 mm)
Ring side clearance:	
Top and second	0.0008 - 0.0018 in (0.02 - 0.045 mm)
Wear limit	0.0059 in (0.15 mm)
Ring gap:	
Top and second	0.0095 - 0.0158 in (0.25 - 0.40 mm)
Wear limit	0.0276 in (0.7 mm)
Oil ring	0.0079 - 0.0354 in (0.2 - 0.9 mm)
Wear limit	0.0433 in (1.1 mm)

Cylinders

Bore I.D.	2.8346 - 2.8352 in (72.00 - 72.015 mm)
Wear limit	2.8386 in (72.1 mm)
Ovality	0.0039 in (0.1 mm)
Wear limit	0.0059 in (0.15 mm)
Cylinder/piston clearance	0.0012 - 0.0028 in (0.03 - 0.47 mm)
Wear limit	0.0059 in (0.15 mm)

Valves and guides

Valve guide clearance	
Inlet	0.004 - 0.0016 in (0.01 - 0.04 mm)
Wear limit	0.0032 in (0.008 mm)
Exhaust	0.0020 - 0.0028 in (0.05 - 0.07 mm)
Wear limit	0.0039 in (0.1 mm)
Valve guide internal diameter	
Inlet and exhaust	0.2598 - 0.2606 in (6.60 - 6.62 mm)
Wear limit	0.2614 in (6.64 mm)
Spring freelength:	
Inner	1.5827 in (40.2 mm)
Wear limit	1.5354 in (39.0 mm)
Outer	1.7225 in (43.75 mm)
Wear limit	1.6732 in (42.5 mm)
Camshaft journal outside diameter:	
Right and left	1.0612 - 1.0619 in (26.954 - 26.970 mm)
Wear limit	1.0595 in (26.91 mm)
Centre	0.9824 - 0.9831 in (24.954 - 24.970 mm)
Wear limit	0.9807 in (24.91 mm)
Cylinder head camshaft bearing:	
Right and left	1.0630 - 1.0638 in (27.0 - 27.021 mm)
Wear limit	1.0623 in (26.98 mm)
Centre	0.9843 - 0.9851 in (25.00 - 25.021 mm)
Wear limit	0.9835 in (24.98 mm)
Cam lift:	
Inlet	0.3512 - 0.3575 in (8.92 - 9.08 mm)
Wear limit	0.3492 in (8.87 mm)
Exhaust	0.3315 - 0.3378 in (8.42 - 8.58 mm)
Wear limit	0.3295 in (8.37 mm)

Big-end bearings

Radial clearance	0.0008 - 0.0017 in (0.02 - 0.044 mm)
Wear limit	0.0032 in (0.08 mm)
Axial clearance	0.0059 - 0.0118 in (0.15 - 0.30 mm)
Wear limit	0.0158 in (0.40 mm)

Crankshaft

Maximum runout	0.0019 in (0.05 mm)

Journal ovality ...	...	...	...	...	...	...	...	0.0002 in (0.005 mm)
Wear limit ...	...	...	...	...	...	...	...	0.0003 in (0.008 mm)
Journal taper ...	...	...	...	...	...	...	...	0.00008 in (0.002 mm)
Wear limit ...	...	...	...	...	...	...	...	0.0002 in (0.004 mm)

Clutch

Type ...	...	...	...	...	...	...	...	Wet, multi-plate
No. of plates:								
Plain ...	...	...	...	...	...	...	...	6
Friction ...	...	...	...	...	...	...	...	8
Friction plate thickness ...	...	...	...	...	...	...	...	0.1347 - 0.1410 in (3.42 - 3.58 mm)
Wear limit ...	...	...	...	...	...	...	...	0.1260 in (3.2 mm)
Plate warpage, max. ...	...	...	...	...	...	...	...	0.0118 in (0.3 mm)
Clutch springs:								
No. of ...	...	...	...	...	...	...	...	6
Free length ...	...	...	...	...	...	...	...	1.3976 in (35.5 mm)
Wear limit ...	...	...	...	...	...	...	...	1.3386 in (34.2 mm)

Oil pumps

Type ...	...	...	...	...	...	...	...	Trochoid
Rotor/bodyside clearance:								
Main pump ...	...	...	...	...	...	...	...	0.0008 - 0.0028 in (0.02 - 0.07 mm)
Wear limit ...	...	...	...	...	...	...	...	0.0047 in (0.12 mm)
Clutch scavenge ...	...	...	...	...	...	...	...	0.0008 - 0.0039 in (0.02 - 0.1 mm)
Wear limit ...	...	...	...	...	...	...	...	0.0047 in (0.12 mm)
Inner rotor/outer rotor clearance:								
Both pumps ...	...	...	...	...	...	...	...	0.0059 in (0.15 mm)
Wear limit ...	...	...	...	...	...	...	...	0.0138 in (0.35 mm)
Rotor/body radial clearance ...	...	...	...	...	...	...	...	0.0059 - 0.0083 in (0.15 - 0.21 mm)
Wear limit ...	...	...	...	...	...	...	...	0.0161 in (0.41 mm)

Gearbox

Type ...	...	...	...	...	...	...	...	5 speed, constant mesh		
Gear ratios										
Reduction 1st gear ...	...	...	...	...	...	...	2.500	Final 1st gear	14.52 : 1
2nd ...	...	...	...	...	...	...	1.708	2nd	9.92 : 1
3rd ...	...	...	...	...	...	...	1.333	3rd	7.74 : 1
4th ...	...	...	...	...	...	...	1.097	4th	6.37 : 1
5th ...	...	...	...	...	...	...	0.939	5th	5.54 : 1
Primary reduction ratio ...	...	...	...	...	...	...	1.708		
Secondary reduction ratio ...	...	...	...	...	...	...	0.825		
Final reduction ratio ...	...	...	...	...	...	...	3.400		
Primary transmission ...	...	...	...	...	...	...	Hy - Vo chain		
Selector fork internal diameter ...	...	...	...	...	...	0.5118 - 0.5125 in (13.00 - 12.018 mm)		
Wear limit ...	...	...	...	...	...	...	0.5134 in (13.04 mm)		
Selector fork rod outside diameter ...	...	...	...	...	0.5105 - 0.5112 in (12.966 - 12.984 mm)		
Wear limit ...	...	...	...	...	...	...	0.5079 in (12.90 mm)		
Fork finger thickness ...	...	...	...	...	...	...	0.2520 - 0.2559 in (6.4 - 6.5 mm)		
Wear limit ...	...	...	...	...	...	...	0.2402 in (6.1 mm)		
Gear change drum outside diameter										
Inside ...	...	...	...	...	...	...	0.4711 - 0.4718 in (11.966 - 11.984 mm)		
Wear limit ...	...	...	...	...	...	...	0.4705 mm (11.95 mm)		
Outside ...	...	...	...	...	...	...	1.4157 - 1.4166 in (35.959 - 35.980 mm)		
Wear limit ...	...	...	...	...	...	...	1.4142 in (35.92 mm)		
Groove width (cam channel) ...	...	...	...	...	...	0.5118 - 0.5125 in (13.0 - 13.018 mm)		
Wear limit ...	...	...	...	...	...	...	0.5134 in (13.04 mm)		
Final gear shaft damper spring:										
Free length ...	...	...	...	...	...	...	4.3661 in (110.9 mm)		
Wear limit ...	...	...	...	...	...	...	3.9370 in (100 mm)		

Torque wrench settings

								lb/ft	kg/cm
Cylinder head nuts ...	...	...	...	...	...	...	...	38 - 41	530 - 570
Cylinder block bolts 6 mm ...	...	...	...	...	...	...	7 - 10	100 - 140
8 mm	...	...	...	...	...	...	...	18 - 21	250 - 290
10 mm	...	...	...	...	...	...	...	24 - 27	330 - 370
Camshaft pulley ...	...	...	...	...	...	...	...	18 - 21	250 - 290
Crankshaft pulley ...	...	...	...	...	...	...	...	24 - 27	330 - 370
Tappet screws ...	...	...	...	...	...	...	...	9 - 12	120 - 160
Alternator centre bolt ...	...	...	...	...	...	...	58 - 65	800 - 900
Clutch centre nut ...	...	...	...	...	...	...	...	27 - 30	380 - 420
Engine mounting bolts 10 mm ...	...	...	...	...	...	22 - 29	300 - 400
12 mm	...	...	...	...	...	...	40 - 43	550 - 600
Main bearing caps ...	...	...	...	...	...	...	...	28 - 30	380 - 420
Big-end bearing caps ...	...	...	...	...	...	...	18 - 21	250 - 290

1 General description

The engine fitted to the Honda Gold Wing is of unusual design, incorporating many features more usual to motor car engine design practice. The engine has four cylinders arranged in two horizontally opposed banks, lying either side of the machine centre line. The aluminium crankcase separates about a vertical plane and houses steel cylinder liners, there being no cylinder barrels as such. The crankshaft runs on three shell type main bearings located by caps bolted to the right-hand crankcase half. The big-end bearings are of a similar type. When the crankcases are separated, the crankshaft and gearbox components which are mounted below, remain in the right-hand casing. The gearbox is placed below the crankshaft, to reduce the overall length of the engine unit; primary drive being transmitted by a Hy-vo chain at the rear of the engine. Each cylinder bank shares a common cylinder head, fitted with two offset valves per combustion chamber, and operated by a single camshaft. The camshafts are driven by two separate toothed bolts from two pulleys mounted in tandem on the extreme front end of the crankshaft. Belt tension is maintained by manually adjustable jockey pulleys, which are tensioned automatically by springs.

Lubrication is of the high pressure, wet sump type, where all the lubricant is contained within the crankcase. Two pumps are fitted, at either end of a single shaft, driven by a duplex roller chain from a sprocket at the rear of the clutch outer drum. The main oil pump is fitted at the front of the engine and supplies oil to the bearing surfaces of the engine via a car type filter mounted on a detachable housing at the extreme front of the engine. Oil returns to the sump by gravity, except for that trapped in the clutch housing, which is returned by the second pump (clutch scavenge pump).

Power from the engine is transmitted to a multiplate clutch and then via a drive shaft to a crown wheel and pinion contained within an aluminium housing at the rear wheel.

In common with many motor car engines, the 1000 cc Gold Wing engine is water cooled. The coolant, which comprises a 50/50 mixture of water and anti-freeze, is circulated around the engine by means of a water pump driven off the forward end of the oil pump drive shaft. The coolant is then passed through a radiator, mounted forward of the engine on the duplex frame down tubes, where it is cooled and returned to the engine. Incorporated in the system is a thermostat, which helps reduce the warming-up period of the engine and regulates the coolant temperature. An electrically driven fan is also fitted to the rear of the radiator, which cuts in automatically when a preset temperature is reached.

2 Operations with engine/gearbox in the frame

Owing to the unusual engine design access to most main components is obscured when the engine is in the frame. The following components can be removed with the engine still in situ:

1 Cylinder heads and valve gear.
2 Timing belts and pulleys.
3 Carburettor assembly.
4 Fuel pump.
5 Contact breaker assembly.
6 Starter motor.
7 Clutch lifting mechanism and plates.

3 Operations with engine/gearbox removed

The engine must be removed for access to and removal of all remaining components, including the following:

1 Clutch.
2 Alternator.
3 Gearchange mechanism.
4 Gearbox components.
5 Crankshaft assembly.
6 Pistons and connecting rods.
7 Clutch outer drum.

4 Method of engine/gearbox removal

The engine and gearbox are built in unit (housed within the same casing) and as such it is necessary to remove the unit complete in order to gain access to either sub-assembly. Separation of the crankcase is achieved after the engine has been removed from the frame and refitting cannot take place until the engine/gear unit is assembled completely.

5 Engine/gearbox unit: removal from the frame

1 Place the machine firmly on its centre stand so that it is standing securely and there is no likelihood that it may fall over. This is extremely important as owing to the weight of the complete machine and the engine, any instability during dismantling will probably be uncontrollable. If possible, place the machine on a raised platform. This will improve accessability and ease engine removal. Again, owing to the weight of the machine, ensure that the platform is sufficiently strong and well supported.
2 Open all three sections of the dummy fuel tank and remove the radiator pressure cap. Place a container of more than 3.4/2.8 US/Imp. quart (3.2 litre) capacity below the front of the engine and remove the coolant drain plug from the lower edge of the water pump outer casing. Allow the coolant to drain. If the coolant is to be reused, ensure that the container is absolutely clean. The container should ideally be of a plastic material as this will not contaminate the coolant. Place a second container below the engine oil drain plug and remove the plug, allowing all the oil to drain. The plug is situated below the oil filter housing. The container must be of more than 3.7/3.1 US/Imp. quart (3.5 litre) capacity.
3 Remove the two frame side covers. Each one is retained at the lower edge by a twist locking screw. Push the screw inwards and twist through 90° to release. Lift the cover away from the bottom and upwards. Prise the rubber boots off the battery terminal and disconnect the positive (+) lead followed by the negative (–) lead. If the machine is to be inoperative for an extended length of time, remove the battery so that it can be serviced and charged at intervals.
4 Loosen the lower hose clip on the radiator upper hose and pull the hose off the thermostat housing union. The radiator lower hose is retained in a similar manner. However, owing to its short length, removal is made easier if the manifold is removed from the water pump outer casing. The manifold is retained by two bolts. Disconnect the radiator fan electrical leads by parting the 'block' connector. Remove the two dome nuts which retain the radiator at the lower mountings. Support the radiator with one hand and remove the similar upper mounting bolts. Turn the forks onto full lock and ease the radiator off the four studs. Take care not to damage the radiator fins.
5 Loosen the silencer clamps at the silencer/exhaust pipe joints. The socket screws which hold the clamps may be caulked, thereby preventing insertion of a suitable socket key. The caulking can be removed using a suitable pointed instrument. Loosen the screws off as much as possible as the clamps are rounded at the front edge to help secure the pipe ends. Loosen and remove the four nuts which retain each exhaust pipe assembly to the cylinder head, and slacken off the silencer bracket/pillion footrest mounting bolts. Tilt the silencer downwards at the front and pull the exhaust pipes from the cylinder heads. The exhaust pipes can now be detached from the silencer. If necessary, the silencer can be supported, using a piece of string tied to a suitable part of the frame.

6   Disconnect the following electrical leads at the 'block' connectors:

Contact breaker and alternator - behind LH side cover.

Engine main harness - from left-hand dummy tank panel.

Also disconnect the starter motor cable at the motor terminal. Pull the socket off the water temperature gauge sensor switch and detach the spark plug suppressor caps. Prise the rubber boot off the oil pressure warning switch and disconnect the wiring lead which is retained by a cross-head screw.

7   Turn the fuel tap to the 'Off' position and loosen the hose clip which secures the fuel feed at the fuel pump. The pipe will pull from position. Disconnect the tachometer drive cable from the gearbox in the fuel pump mounting casing by removing the inner flange retaining bolt. Replace the bolt after pulling the cable from position.

8   Disconnect the clutch cable at the handlebar control lever and pull it through slightly. Remove the clutch adjustment cover from the rear of the engine and disconnect the inner cable from the operating arm. Unscrew the cable adjuster and pull the cable from position. Access to the clutch lifting mechanism is difficult. If necessary, cable removal can take place after the engine is partially removed from the frame. Disconnect the throttle cables at the twist grip. The two halves of the grip clamp must be loosened by removing the two retaining screws. Loosen the locknuts on the throttle cables at the carburettor end. Remove the outer cable of the rearmost cable from the anchor bracket and detach the inner cable nipple from the throttle pulley. Repeat for the final cable. Access to the throttle cables is also difficult and this operation too can be accomplished, if necessary, when the engine is partially removed from the frame. Disconnect the choke cable at the carburettors. The outer cable is clamped to the holder bracket by a screw and clamp.

9   Remove the horn from the right-hand top tube where it is retained by two bolts. Disconnect the two 'spade' connectors. Remove the air filter cover and air filter which is retained by a wing nut. Remove the two screws which retain the filter box to the carburettor air box.

10 Prise the rubber gaiter off the final drive shaft where it locates with the rear of the engine. Using a pair of externally · opening 90° cranked jaw circlip pliers, remove the circlip which retains the forward 'knuckle' of the drive shaft universal joint to the engine shaft. This operation requires care, as the circlip is partially obscured. After removal of the circlip, prise the complete joint backwards until it leaves the splines of the engine shaft.

11 Because of the great weight of the engine unit and the small amount of room available for manoeuvering, an hydraulic trolley jack should be used to support the engine as it is removed from the frame. Although it may be possible in practice to lift the engine clear with the aid of two or three assistants, the likelihood of damage to the engine is very great and should be avoided at all costs. The complete engine removal operation is as follows:

Place the trolley jack at right angles to the engine centre line, on the left-hand side of the machine. If the machine is supported on a raised platform, an extension to the platform should be arranged upon which the jack can stand and be pulled outwards, bearing the weight of the engine. Place a suitable plank of wood below the engine sump and position the jack so that the engine is slightly supported. Try and position the jack below the engine so that the engine will be balanced on removal. This is largely a matter of guesswork.

12 Remove the four bolts which hold the fan shroud to the upper front engine mounting lugs and to the frame. Tilt the shroud forwards at the upper edge and unclip the wiring leads. The shroud can now be removed. Disconnect the push-fit lead to the electric fan sensor switch. Remove the front engine mounting bolt and the lower rear mounting bolts from both the right and left-hand side. Adjust the engine height by means of the jack as these bolts are removed. Detach the removable left-hand engine cradle frame member. The member is retained by two studs and dome nuts at the front, and by a stud and bolt at the rear. Raise the engine slightly and remove the rear upper mounting cross-bolt, followed by the mounting brackets. Note the earth lead retained behind the left-hand bracket.

13 The engine is now free for removal. Have the engine supported by two assistants and ensuring that it is at height so that it will not foul the frame, gently wheel it outwards to the left. The air filter box will have to be supported to clear the carburettor assembly. If the throttle cables and clutch control cable have not been disconnected this should be accomplished before the engine is moved further. Check that the electrical leads are completely disconnected and move the engine outwards until it is completely clear of the frame. The engine must now be lifted clear of the jack and placed securely on the workshop floor or workbench.

---

6   Dismantling the engine/gearbox unit: general

---

1   Before commencing work on the engine, the external surfaces must be cleaned thoroughly. A motorcycle engine has very little protection from road dirt, which will sooner or later find its way into the dismantled engine if this simple precaution is not observed.

2   One of the proprietary engine cleaning compounds such as Gunk or Jizer can be used to good effect, especially if the compound is allowed to penetrate the film of oil and grease before it is washed away. When washing down, make sure that water cannot enter the carburettors or the electrical system, particularly if these parts are now more exposed.

3   Never use force to remove any stubborn part, unless mention is made of this requirement in the text. There is invariably good reason why a part is difficult to remove, often because the dismantling operation has been tackled in the wrong sequence.

4   Dismantling will be made easier if a simple engine stand is constructed that will correspond with the engine mounting points. This arrangement will permit the complete unit to be clamped rigidly to the workbench, leaving both hands free for the dismantling operation.

5   Most operations on the engine can be carried out using normal hand tools. Under certain circumstances special tools will be required either because the use of an incorrect tool may damage a component or because the accuracy required can only be maintained using a particular tool or instrument. Apart from normal hand tools a number of additional tools should be acquired, as follows:

Torque wrench
Impact screwdriver
Pair of internal opening circlips
Pair of external opening circlips (straight jaws)

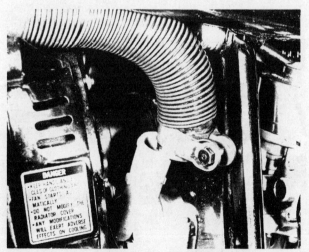

5.4a Disconnect radiator top hose at manifold

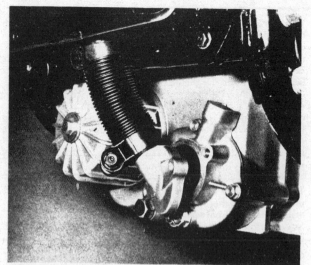

5.4b Detach manifold to remove lower hose

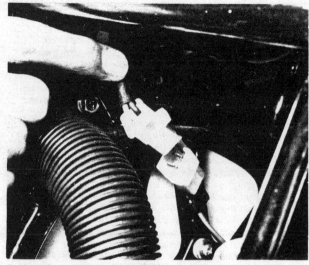

5.4c Disconnect fan leads at block connector

5.4d Remove upper and lower bolts and ...

5.4e ... lift radiator away carefully

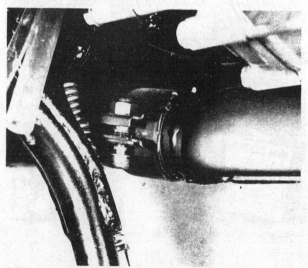

5.5a Loosen silencer at pipe joint and ...

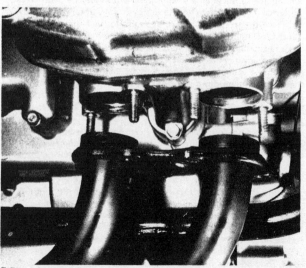

5.5b ... free the pipe ends from exhaust port

5.5c The exhaust pipe can be pulled free

5.6a Disconnect main harness from electrical panel

5.6b Prise off boot and detach starter motor cable

5.6c Disconnect engine sub harness at sensor switches

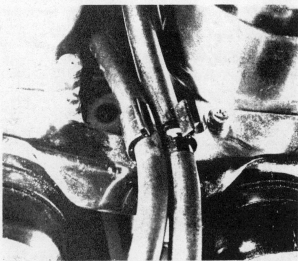
5.6d Unclip HT leads and disconnect at plugs

5.6e Oil pressure switch lead held by single screw

5.7 Disconnect fuel line to pump union

5.8a Disconnect the throttle cables followed ...

5.8b ... by the choke cable

5.9 Detach horn from top tube

5.10a Prise the rubber gaiter from position and remove circlip

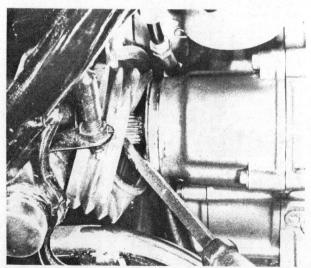

5.10b Force joint back along splined shaft

5.12a Tilt fan shroud to remove and ...

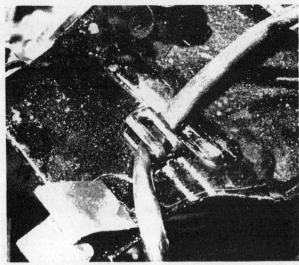

5.12b ...disconnect leads at clips on rear of shroud

5.12c Engine retained by lower cross bolt at front and ...

5.12d ...side bolts on rear at either side

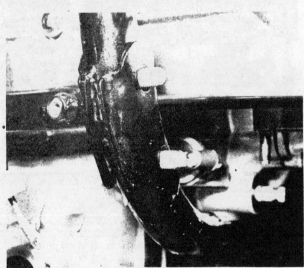

5.12e Detach the subframe held by two bolts at the front ...

5.12f ...and by a bolt and nut and stud at rear

5.12g Remove upper mountings from either side

7 Dismantling the engine unit: removing the carburettors

1   The carburettors can be removed as a single unit, further
dismantling only being necessary if the carburettors require
attention. Remove the two dome nuts which retain each
carburettor manifold to the cylinder head. Lift the complete unit
upwards, off the retaining studs.

8 Dismantling the engine unit: removing the thermostat housing

1   Remove the three screws which hold the thermostat housing
to the crankcase, and the two screws which retain each transfer
manifold. Lift the complete unit from position. Note the hollow
dowel and 'O' ring which locate the housing. Pull the transfer
pipes out of the housing and their manifolds. Note the four 'O'
rings. If required, the hose manifold can be detached from the
housing, after removing the two bolts, and the thermostat lifted
from position.

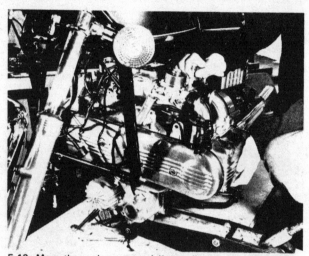

5.13a Move the engine out partially, to disconnect controls

7.1 Lift carburettor assembly away complete

5.13b Support engine, carefully balanced on jack

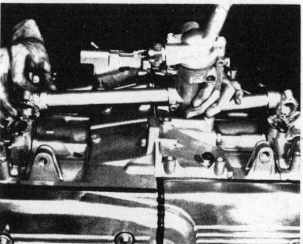

8.1 Thermostat assembly will lift away as complete unit

9 Dismantling the engine unit: removing the timing belts

1 Remove the left-hand timing belt cover followed by the right-hand cover. Both covers are held by two bolts. Slacken off the jockey pulley bolts.

2 Ease the outermost (right-hand) toothed belt from position on the pulleys. This is most easily accomplished if the crankshaft is rotated by means of a spanner on the pulley bolt. Remove the inner belt. Considerable care should be exercised when removing the belts as they can easily become damaged. The belts are manufactured from synthetic rubber strengthened by glass fibre strands, and bending to a radius of less than 1 inch (25 mm) or scoring with a screwdriver during removal, will considerably shorten their life. Avoid contact between the belts and engine oil or grease as the lubricant may swell the rubber, causing deterioration and maladjustment of valve timing.

3 Before removing the timing belts, mark each with a piece of tape to indicate the original direction of travel. If a belt is subsequently fitted so that the direction of travel is reversed, belt wear will be accelerated.

4 Remove the drive pulley centre bolt. The bolt can be loosened by using the weight of the reciprocating parts to prevent the crankshaft turning. Pull the two pulleys from position together with the keeper plates. Remove the centre bolts from the two driven pulleys and draw the pulleys from position. Note the Woodruff key in each camshaft, which should be removed using a suitable screwdriver. Before removing the pulleys, check that both are marked on the outer face so that no confusion will arise on re-assembly. The pulleys can be prevented from rotating while the centre bolt is removed, by placing a tyre lever through the spokes, abuting against one of the bolt heads which lie behind the pulley. **Warning:** Do not allow the camshafts or the crankshaft to rotate independently of each other as there is a danger that a piston may make contact with an open valve.

5 Remove the inner case extension from each cylinder head. Each is retained by two bolts.

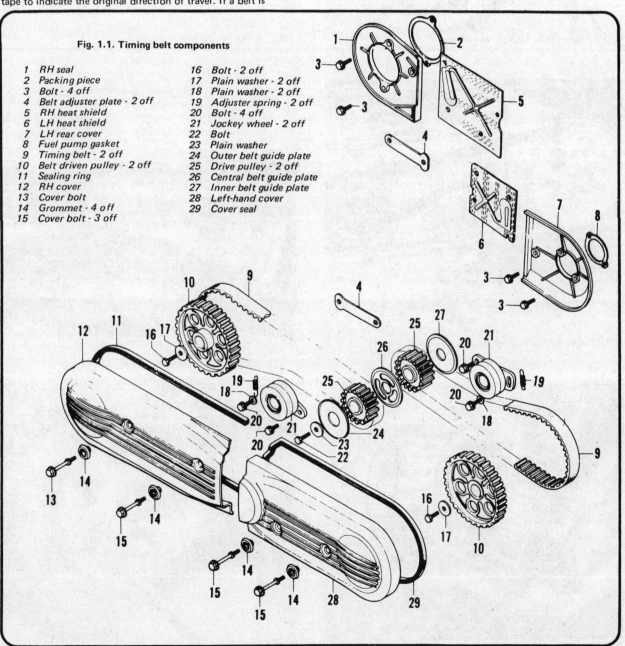

Fig. 1.1. Timing belt components

1	RH seal	16	Bolt - 2 off
2	Packing piece	17	Plain washer - 2 off
3	Bolt - 4 off	18	Plain washer - 2 off
4	Belt adjuster plate - 2 off	19	Adjuster spring - 2 off
5	RH heat shield	20	Bolt - 4 off
6	LH heat shield	21	Jockey wheel - 2 off
7	LH rear cover	22	Bolt
8	Fuel pump gasket	23	Plain washer
9	Timing belt - 2 off	24	Outer belt guide plate
10	Belt driven pulley - 2 off	25	Drive pulley - 2 off
11	Sealing ring	26	Central belt guide plate
12	RH cover	27	Inner belt guide plate
13	Cover bolt	28	Left-hand cover
14	Grommet - 4 off	29	Cover seal
15	Cover bolt - 3 off		

9.1a Remove the left hand cover before ...

9.1b ...detaching the right-hand cover

9.1c Slacken jockey pulleys to remove belts

9.3 Mark direction of belt travel before removal

9.4a Pull drive pulleys and guide plates off crankshaft

9.4b Lock pulley to loosen centre bolt

9.4c Note markings on pulleys before removal

9.5 Detach inner case extensions, held by two bolts

10 Dismantling the engine unit: removing the cylinder heads

1   Loosen and remove the two flange bolts and detach the fuel pump and tachometer drive assembly as a complete unit. The contact breaker assembly and housing which occupies a similar position on the rear of the left-hand cylinder head need not be removed at this stage unless the camshaft is to be removed or the unit itself requires attention.
2   If required, remove the contact breaker assembly as follows: Release the low tension cable by prising the wiring clamp apart. Remove the two screws which hold the housing to the cylinder head. The housing, complete with the contact breaker assembly can be pulled from position. Remove the cam bolt and draw the Automatic Timing Unit (ATU) from position on the end of the camshaft. Note and remove the Woodruff key.
3   The cylinder heads should be removed individually using an identical procedure, as follows:
    Unscrew the cylinder head/cam cover retaining bolts and lift off the cover. Place a small container or old rag below the cylinder head to catch the small amount of oil trapped in the cover. When removing the cover, take care not to damage the rubber sealing ring.

4   Loosen the seven cylinder head retaining bolts evenly and in a diagonal sequence. This will help prevent distortion. Note that in addition to the six 10 mm bolts, an extra 6 mm bolt is fitted on the lower flange of the cylinder head. The cylinder head gasket is of the impregnated type and consequently after a certain amount of service the head will become bonded to the cylinder barrel. A rawhide mallet may be used to free it, or in extreme cases, a block of soft wood placed on suitable portions of the head and tapped with a hammer may be utilised. Have an assistant support the cylinder head so that it does not fall free and become damaged. After removal of the head, pull the two hollow dowels from position together with the oil feed control connector. Note the two 'O' rings on the connector.

11 Dismantling the engine unit: removing the clutch

1   Before attending to the clutch, remove the breather box and hoses from above the rear engine cover. The box is retained by two bolts. Also remove the kickstart arm and crank. Both components are retained on their respective shafts by a circlip.
2   Remove the clutch cover, which is retained by six bolts and two nuts and studs. Note the position of the wiring lead clamp retained by the lower of the two nuts. Unscrew the six spring tension bolts which retain the clutch pressure plate. The pressure plate will lift out. The clutch centre boss is retained by a special centre nut which requires a peg spanner. A suitable tool can be fabricated from a short length of thick walled pipe (plumbers pipe). Relieve the end of the pipe so that two or four pegs are formed which will engage securely with the special nut. Bend down the ears of the tab washer before attempting to loosen the nut. To prevent the clutch centre rotating replace two or more clutch springs and bolts using suitable washers to take the place of the pressure plate. Sufficient pressure will be given to lock the two parts of the clutch together. Remove the inspection cap from the alternator casing and apply a spanner to the rotor nut. The clutch centre nut can now be loosened.
3   Remove the refitted clutch springs and bolts and pull out the clutch centre followed by the clutch plates. Carefully note the sequence of plates to aid replacement, particularly that the plates are fitted in two separate sections, separated by a special double thickness plate. Finally, remove the internal splined washer. The clutch outer drum must remain in position until the engine rear cover has been removed.

12 Dismantling the engine unit: removing the starter motor

1   The starter motor is retained by two bolts passing through lugs on the motor body. Remove the bolts and using a rawhide mallet, gently tap the starter motor from the casing. The sprocket, with which the starter splined shaft engages, will remain in situ within the rear cover, still meshed with the chain.

13 Dismantling the engine unit: removing the rear cover and clutch outer drum

1   Loosen the eleven bolts which retain the rear cover. Before removal, note the wiring lead clip retained by the centre bolt on the left-hand side. The case is now free for removal. A rawhide mallet may be required to ease the case off the two hollow locating dowels. When removing the case, push the final drive splined shaft inwards, so that it is not disturbed.
2   Remove the bolt and washer from the end of the oil pump drive shaft to free the oil pump sprocket, and remove the clutch outer drum retaining circlip. Ease the oil pump driven sprocket and the clutch outer drum off their respective shafts simultaneously. The sprockets can then be detached from the duplex chain.

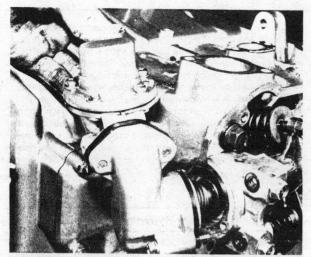
10.1 Detach fuel pump/tachometer drive as unit

10.3 Remove the rocker covers from both cylinders and ...

10.4a ... unscrew the six rocker support bolts to ...

10.4b ... detach support and remove camshaft (if required)

10.4c Use block of wood to remove heads if required

10.4d Note oil feed nozzle, and remove

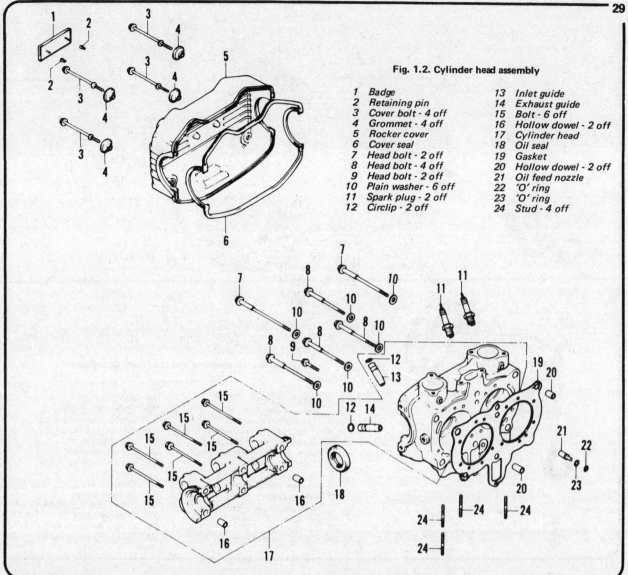

Fig. 1.2. Cylinder head assembly

1 Badge	13 Inlet guide
2 Retaining pin	14 Exhaust guide
3 Cover bolt - 4 off	15 Bolt - 6 off
4 Grommet - 4 off	16 Hollow dowel - 2 off
5 Rocker cover	17 Cylinder head
6 Cover seal	18 Oil seal
7 Head bolt - 2 off	19 Gasket
8 Head bolt - 4 off	20 Hollow dowel - 2 off
9 Head bolt - 2 off	21 Oil feed nozzle
10 Plain washer - 6 off	22 'O' ring
11 Spark plug - 2 off	23 'O' ring
12 Circlip - 2 off	24 Stud - 4 off

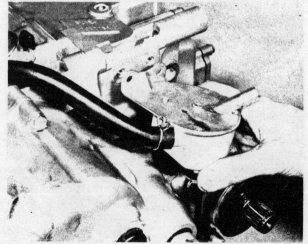

11.1a Remove the breather box, held by two bolts

11.1b Detach kickstart joint, retained by circlip ...

Fig. 1.3. Clutch assembly and final drive output shaft

1	Oil drain plate	15	Ball bearing
2	Journal ball bearing	16	Clutch lifter plate
3	Final driven gear - 33T	17	Special curved washer
4	Final drive output shaft	18	Special peg nut
5	Cam damper - fixed piece	19	Tab washer
6	Cam damper - moving piece	20	Clutch spring
7	Damper spring	21	Clutch friction plate - special
8	Spring retaining collet - 2 off	22	Clutch plain plate - 6 off
9	Journal ball bearing	23	Damper plate
10	Spring seat	24	Clutch friction plate - 7 off
11	Internal splined washer	25	Pressure plate
12	Clutch centre boss	26	Clutch outer drum
13	Spring bolt - 6 off	27	Oil pump drive chain
14	Clutch lifter piece		

11.1c ...as is the kickstart crank piece

11.2a Unscrew six bolts to free lifting plate and springs

11.2b Refit two springs to loosen centre nut with peg spanner

11.3 Pull clutch centre out together with plates

13.1 Remove rear engine cover

13.2a Remove the large special washer and ...

13.2b ...the clutch retaining circlip

13.2c Pull clutch outer drum off shaft together with pump drive

14 Dismantling the engine unit: removing the alternator rotor and starter clutch assembly

1   Loosen and remove the alternator rotor centre nut. The rotor and starter clutch assembly, to which it is attached, can be drawn off the splined shaft. It is likely that the assembly will be tight on the splines, in which case an extractor must be used for removal. A standard two or three legged sprocket puller may be employed. Do not try prising the assembly from position using screwdrivers or other tools as levers. The mating surfaces of the casing and the rear face of the starter clutch will probably suffer damage. Be prepared to take the weight of the alternator rotor as it leaves the shaft. It is a heavy component weighing 9 lbs.

2   Remove the splined washer from the alternator shaft. Loosen the chain guide retainer bolt and rotate the guide so that it clears the chain. Pull the sprocket off the shaft together with the chain and starter drive sprocket.

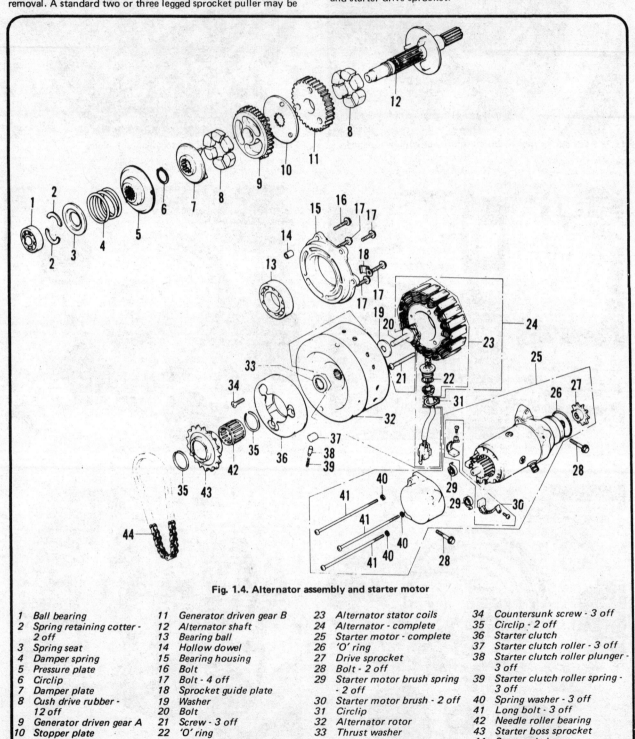

Fig. 1.4. Alternator assembly and starter motor

1 Ball bearing	11 Generator driven gear B	23 Alternator stator coils	34 Countersunk screw - 3 off
2 Spring retaining cotter - 2 off	12 Alternator shaft	24 Alternator - complete	35 Circlip - 2 off
3 Spring seat	13 Bearing ball	25 Starter motor - complete	36 Starter clutch
4 Damper spring	14 Hollow dowel	26 'O' ring	37 Starter clutch roller - 3 off
5 Pressure plate	15 Bearing housing	27 Drive sprocket	38 Starter clutch roller plunger - 3 off
6 Circlip	16 Bolt	28 Bolt - 2 off	39 Starter clutch roller spring - 3 off
7 Damper plate	17 Bolt - 4 off	29 Starter motor brush spring - 2 off	40 Spring washer - 3 off
8 Cush drive rubber - 12 off	18 Sprocket guide plate	30 Starter motor brush - 2 off	41 Long bolt - 3 off
9 Generator driven gear A	19 Washer	31 Circlip	42 Needle roller bearing
10 Stopper plate	20 Bolt	32 Alternator rotor	43 Starter boss sprocket
	21 Screw - 3 off	33 Thrust washer	44 Starter chain
	22 'O' ring		

14.1a Remove alternator rotor centre bolt and ...

14.1b ... pull off alternator rotor complete with starter clutch

14.1c Loosen clutch boss sprocket guide plate

15 Dismantling the engine unit: removal of final drive engine shaft

1   Remove the cover which lies to the rear of the right-hand rear cylinder. The cover is retained by four bolts. The drive shaft will pull from position as a complete unit, leaving the final drive double gear in position. The double gear is removable through the opening in the case.

16 Dismantling the engine unit: removing the oil filter, water pump and front engine cover

1   Unscrew the oil filter housing centre bolt and remove the housing, complete with filter.
2   Remove the four screws which retain the water pump outer casing. The pump casing is located on two dowels and fitted with a self-sealing gasket; it may be tight and require easing with a rawhide mallet. Remove the nine screws which retain the front cover and lift the cover from position. Remove the cover gasket, noting the four 'O' rings and two collars.
3   The water pump is retained in the front cover, from the inside by three bolts. After removal of the bolts, tap the pump body gently until it leaves the casing. Do not tap the pump shaft as damage may result. Note the two 'O' rings which fit in locating grooves around the double diameter periphery of the pump body.

17 Dismantling the engine unit: separating the crankcase halves

1   Remove the three crankcase half holding bolts from the right-hand side of the engine. One bolt is positioned at the upper rear corner and two at the lower rear corner. Place the engine so that the right-hand cylinder block mating surface rests securely on the workbench. Disengage the spring loaded change pawl, which pivots on the gearchange arm, from the change drum pins. Tie the pawl back to the change arm so that it will not snag any part of the gear change mechanism in the right-hand casing.
2   Loosen the nineteen left-hand crankcase bolts evenly and in a diagonal sequence. Remove all the bolts. Separation of the crankcase halves requires two people. One person to lift the left-hand casing upwards and the second person to support the pistons, preventing them from falling free and becoming damaged as they leave the cylinder bores. Use a rawhide mallet to loosen the cases. Lift the left-hand case upwards so that it remains square in all planes and the pistons do not tie in the cylinder bores. Broken piston rings will fall into the right-hand crankcase from where they should be retrieved immediately. Place a clean rag over the upper edge of the right-hand casing upon which the two pistons may rest.

15.1 Pull drive shaft out to release double drive gear

16.2a Remove water pump outer casing followed ...

16.2b ...by the engine front cover

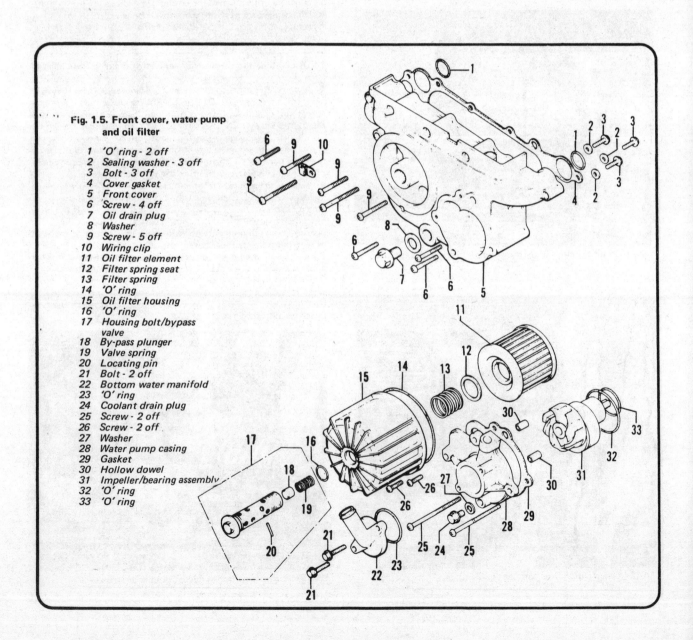

Fig. 1.5. Front cover, water pump and oil filter

1 'O' ring - 2 off
2 Sealing washer - 3 off
3 Bolt - 3 off
4 Cover gasket
5 Front cover
6 Screw - 4 off
7 Oil drain plug
8 Washer
9 Screw - 5 off
10 Wiring clip
11 Oil filter element
12 Filter spring seat
13 Filter spring
14 'O' ring
15 Oil filter housing
16 'O' ring
17 Housing bolt/bypass valve
18 By-pass plunger
19 Valve spring
20 Locating pin
21 Bolt - 2 off
22 Bottom water manifold
23 'O' ring
24 Coolant drain plug
25 Screw - 2 off
26 Screw - 2 off
27 Washer
28 Water pump casing
29 Gasket
30 Hollow dowel
31 Impeller/bearing assembly
32 'O' ring
33 'O' ring

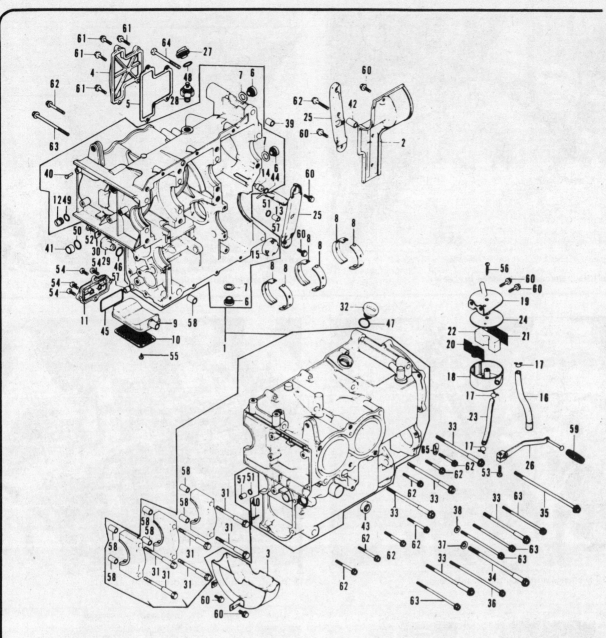

Fig. 1.6. Crankcase assembly

1 Crankcases - complete	17 Hose clip - 3 off	34 Bolt	50 'O' ring
2 Oil catch plate	18 Breather box	35 Bolt	51 'O' ring - 2 off
3 Primary chain oil catch plate	19 Breather box cover	36 Bolt	52 Bolt
4 Final gear cover	20 Gauze	37 Washer	53 Bolt
5 Final gear cover gasket	21 Gauze	38 Washer	54 Screw - 4 off
6 Blanking plug - 3 off	22 Filter element	39 Hollow dowel	55 Screw - 3 off
7 Sealing washer - 3 off	23 Breather hose	40 Pin	56 Screw
8 Main bearing shell - 6 off	24 Packing piece	41 Hollow dowel	57 Hollow dowel - 4 off
9 Oil strainer case	25 RH primary chain guide	42 Hollow dowel	58 Hollow dowel - 7 off
10 Oil strainer gauze	26 Gear change pedal	43 Oil seal	59 Gear change pedal rubber
11 Oil strainer cover	27 Oil filler cap	44 'O' ring	60 Bolt - 9 off (11 off - '76 model)
12 Oil feed dowel	28 Oil pressure switch	45 Oil strainer sealing ring	61 Bolt - 4 off
13 Oil feed nozzle	29 Neutral gear switch	46 'O' ring	62 Bolt - 9 off
14 Oil feed orifice	30 Switch retainer plate	47 'O' ring	63 Bolt - 5 off
15 Oil level sight glass	31 Bolt - 6 off	48 'O' ring	64 Bolt
16 Breather hose	32 Timing mark plug	49 'O' ring	65 Wiring clip
	33 Bolt - 5 off		

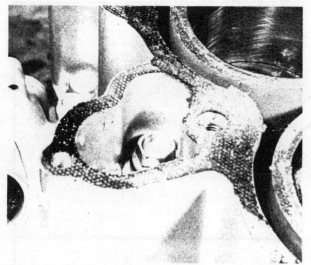

17.1a Note crankcase bolt hidden in recess

18.2a Lift mainshaft to remove oil thrower plate

17.1b Tie change arm back to prevent snagging

18.2b Tilt assembly to remove mainshaft and ...

17.2 Support pistons as casing is lifted upwards

18.2c ...then disengage the primary drive gear

18 Dismantling the engine unit: removing the crankshaft, pistons and gearbox components

1   Check that the main bearing caps and big-end bearing caps are marked in relation to their positions. If they have not been marked, this should be accomplished now, using a punch or scribe, so that no confusion will arise on replacement as to the correct positions. Do not obliterate the scribed letters and numbers already marked, as these are code numbers for journal and shell sizes, which will be required when ordering spare parts.
2   Remove the three bolts which retain the primary driven gear oil throw shield. Lift the primary driven shaft (gearbox main-shaft) upwards slightly as a complete unit so that the oil throw shield can be slid round the primary driven gear and removed. Tilt the shaft upwards at the forward end and pull the shaft complete with gear pinions from position in the primary driven gear. The primary driven gear can now be disengaged from the Hy-vo primary drive chain.
3   Remove the big-end bearing cap nuts from the two right-hand connecting rods, rotating the crankshaft as necessary. Remove the bearing caps and push the connecting rods away from the big-end bearing journals. Remove the main bearing cap bolts followed by the caps. If a bearing shell falls out of either a big-end cap or main bearing cap it should be replaced immediately to avoid confusion. Grasp the left-hand pistons and the Hy-vo chain and lift them upwards complete with the crankshaft. Take care not to damage the piston rings when lifting.
4   Replace the bearing caps in their original positions to aid later identification.
5   Lift the oil gauze filter box from position in the crankcase. Remove the nut from the gear change drum stopper arm pivot shaft and remove the two stopper arms and the spacer washers. Carefully note the sequence of washers and spacers and the

position of the arms before removal. Remove the pivot bolt which retains the change drum stopper inner claw assembly. If care is taken, the claw can be manoeuvered off the end of the change drum. Grasp the end of the retainer pin which passes vertically through the crankcase and locates the selector fork rod. The retaining pin will pull out of the case. Note the relative positions of the selector forks and push the selector rod out of position in the case so that all three selector forks are freed. The selector fork rod has a slotted end. A screwdriver can be used to rotate the rod if difficulty in removal is experienced.
6   Remove the neutral stopper switch from the outside of the casing. The switch is a push-fit and is retained by a tongue held by a single bolt. The gear change drum can now be pulled out of the casing.
7   Using an 'impact' screwdriver remove the two countersunk screws which retain the blind layshaft bearing housing. The bearing housing is a push-fit and can be removed, complete with the bearing. Pull the 5th gear pinion off the layshaft and remove it through the bearing orifice in the gearbox wall. Slide the complete layshaft assembly out of the rearmost bearing and lift it out of the casing.
8   Remove the various bearing half clips and locating dowels from the edge of the crankcase. Position the crankcase so that access can be gained to the pistons. Push each connecting rod up the bore so that the pistons become free. If the engine has covered a substantial mileage a wear lip will have formed at the top of the bore. This marks the upper limit of piston top ring travel. This lip MUST be removed with the aid of a cylinder bore lapper before the pistons are removed. If the lip is not removed the piston rings will certainly break, as the piston is removed. Removal of the pistons from the connecting rods should not take place at this stage due to the reasons given in Section 25 of this Chapter.

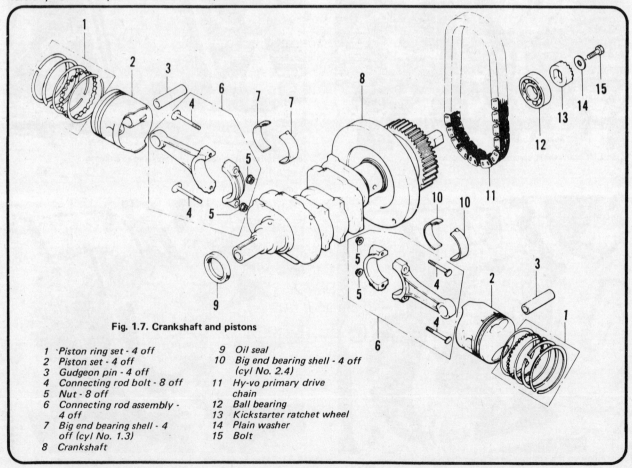

Fig. 1.7. Crankshaft and pistons

1 Piston ring set - 4 off
2 Piston set - 4 off
3 Gudgeon pin - 4 off
4 Connecting rod bolt - 8 off
5 Nut - 8 off
6 Connecting rod assembly - 4 off
7 Big end bearing shell - 4 off (cyl No. 1.3)
8 Crankshaft
9 Oil seal
10 Big end bearing shell - 4 off (cyl No. 2.4)
11 Hy-vo primary drive chain
12 Ball bearing
13 Kickstarter ratchet wheel
14 Plain washer
15 Bolt

18.3a Disconnect right hand connecting rods from crankshaft

18.3b Mark and remove main bearing caps and shells

18.3c Lift crankshaft assembly from position

18.5a Remove the gauze oil trap from crankcase

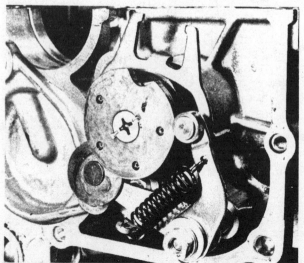
18.5b Remove the stopper arms from the pivot shaft and ...

18.5c ...Manoeuvre the change claw from the change drum

18.5d Pull the selector rod locating pin from the casing and ...

18.5e ...draw the rod out to free the selector forks

18.6a Remove the neutral stopper switch and ...

18.6b ...draw the change drum from the gearbox

18.7a Pull out the layshaft bearing housing to ...

18.7b ...gain access to the 5th gear pinion

18.8a Remove layshaft by tilting

18.8b Prise bearing clips from the gearbox wall

18.8c Remove the right hand pistons from the bores ...

19 Dismantling the engine unit: removing the oil pumps, alternator shaft and gear change shaft

1   All the components listed in the Section heading are located in the left-hand crankcase half.

2   Remove the Hy-vo chain guide which is retained by two bolts, which also retain the oil catch plate. Detach the oil feed trough from the crankcase wall. The alternator shaft, complete with pinion and shock absorber, is retained by a detachable bearing housing. Remove the six retaining bolts and pull the housing complete with shaft from position. Remove the primary chain feed nozzle, which is a push-fit in the case.

3   The oil pumps are fitted at either end of the crankcase, being driven by a shaft which passes through and between the two. Detach the clutch scavenge pump, which is the smallest of the two and retained by three bolts. Pull the pump off the shaft. The main pump can be removed in a similar manner, together with the drive shaft, which will remain connected. Note the two hollow dowels on which the pump locates.

4   Loosen the gear change shaft ball arm bolt after knocking down the ears of the tab washer. Pull the shaft out of the casing, noting the centraliser spring and spring collar.

5   Remove the remaining plate retaining bolt and lift the oil catch plate from position.

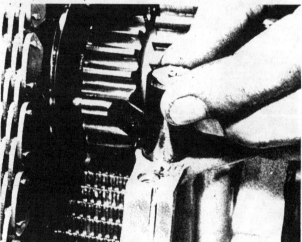

19.1 Pull out the primary chain oil feed nozzle

19.2a Remove the 'HY-VO' chain guide

19.2b Detach the oil catch trough ...

19.2c ...note that the nuts are 'capped'

19.2d Unscrew six bolts to withdraw alternator gear

19.3a Clutch scavenge pump held by three bolts

19.3b Main pump will pull out, attached to drive shaft

19.4 Knock down tab washer before loosening bolt

19.5 Lift oil catch plate out after extracting pump and shaft

20 Examination and renovation: general

1 Before examining the parts of the dismantled engine unit for wear, it is essential that they should be cleaned thoroughly. Use a paraffin/petrol mix to remove all traces of old oil and sludge that may have accumulated within the engine.
2 Examine the crankcase castings for cracks or other signs of damage. If a crack is discovered, it will require professional repair, or renewal.
3 Carefully examine each part to determine the extent of wear, checking with the tolerance figures listed in the Specifications Section of this Chapter. If there is any question of doubt, play safe and renew,
4 Use a clean lint free rag for cleaning and drying the components. This will obviate the risk of small particles obstructing the internal oilways, causing lubrication failure.
5 Various instruments for measuring wear are required including a vernier gauge or external micrometer and a set of standard feeler gauges. Honda recommend the use of 'plastigauge' for measuring radial clearances between working surfaces such as shell bearings and their journals. 'Plastigauge' consists of a plastic sheet supplied in narrow strips upon which is marked specific units of measurement. A small length of plastigauge is placed between the two surfaces, the clearance of which is to be measured. The surfaces are bolted up in their normal working positions and then separated. The amount of compression that the gauge material is subjected to and the resultant lengthening of the distances between the pre-marked units of length, indicates the exact clearance. If 'plastigauge' is not available both an internal and external micrometer will be required to check wear limits. Additionally, although not absolutely necessary, a dial gauge and mounting bracket is invaluable for accurate measurement of end float, and play between components of very low diameter bores - where a micrometer cannot reach.

21 Big-end bearings: examination and renovation

1 Big-end failure is invariably indicated by a pronounced knock from the crankcase. The knock will become progressively worse, accompanied by vibration. It is essential that the bearings are renewed as soon as possible since the oil pressure will be reduced and damage caused to other parts of the engine.
2 Bearing wear at the big-ends can only be really accurately assessed after separating the crankcase halves. A rough indication can, however, be obtained after removal of the cylinder head in the following manner. Starting on No. one (1) cylinder, rotate the engine so that the piston reached TDC and then starts to

move down the bore. Stop the crankshaft. In this position the big-end journal is pulling away from the connecting rod. Using two thumbs press down sharply on the piston crown. Any movement of the piston which is not accompanied by movement of the crankshaft indicates that wear has developed either at the big-end or between the working surfaces of the gudgeon pin and piston bosses. In either case the crankcases will have to be separated for further examination and work.
3 The big-ends have shell type bearings. Examine the bearing surfaces after removing the bearing caps; it is not necessary to remove the shell. If the bearings are badly scuffed or scored the shells will have to be renewed. Always renew bearings as a complete set. If the shell surfaces are excessively scuffed or have 'picked-up' and the journal on that bearing is 'blued' or discoloured, it may be an indication of lubrication failure. The functioning of the lubrication system MUST be checked before engine reassembly in such a case.
4 If the condition of the bearings appear to be satisfactory, check that the clearance between each bearing and the journal is within the recommended limit as follows:

Standard clearance	0.0008 - 0.0017 in
	(0.02 - 0.044 mm)
Wear limit	0.0032 in
	(0.08 mm)

The clearance can be assessed by measuring the internal diameter of the shell bearing and the outside diameter of the big-end journal and subtracting the second figure obtained from the first. 'Plastigauge' can also be used in the following manner. Cut a short length of 'Plastigauge' and place it on the journal so that the pre-marked indexing runs axially. Bolt up the connecting rod and bearing cap to the recommended torque of 18 - 21 lb ft (250 - 290 kg cm). Do not rotate the bearing. Separate the two bearing halves and remove and inspect the 'Plastigauge'. This will indicate the amount as clearance reading of the 'Plastigauge' will depend on the type used. See manufacturer's instructions. The clearance may also be ascertained by direct measurement of the thickness of the 'Plastigauge', using a micrometer.
5 Bearing selection, when renewing the bearings, should be made by referring to the connecting rod code; a number stamped on the upper machined side of the connecting rod, and the big-end journal code, which takes the form of letters scribed on the adjacent crankshaft webs. The letters and numbers should be cross-referred to the accompanying selection chart. The letter stamped on each connecting rod is the weight code.
6 When fitting new bearings, ensure that they are positioned correctly and that the tongues on the end of each shell locate with the recesses in the connecting rod or bearing cap. Also check the clearance on each bearing to ensure that selection is accurate. It is considered good practice to renew all bearing shells when an engine is stripped down, irrespective of their condition. Shell bearings are relatively inexpensive compared with the cost of subsequent dismantling to replace bearings that have failed prematurely.

ROD BEARING SELECTION

CONNECTING ROD SIZE CODES				
	3	BROWN	BLACK	BLUE
	2	GREEN	BROWN	BLACK
	1	YELLOW	GREEN	BROWN
		A	B	C
		CRANKSHAFT CONNECTING ROD JOURNAL SIZE CODES		

Fig. 1.8. Connecting rod bearing selection chart

21.2 Inspect bearing surfaces for scoring

21.6a Ensure correct fitting of new shells

22 Crankshaft and main bearings: examination and renovation

1   Examine the main bearing shells and check the clearances, using the procedure as described for big-end bearings. The correct clearances are as follows:

Standard clearance	*0.0008 - 0.0017 in (0.02 - 0.044 mm)*
Wear limit	*0.0032 in (0.08 mm)*

2   Measure the main bearing and big-end bearing journals for ovality by measuring each journal in several radial positions, with a micrometer. If any journal is worn beyond the service limit, the crankshaft must be renewed.

Standard ovality	*0.0002 in (0.005 mm)*
Wear limit	*0.0003 in (0.008 mm)*

Check each journal for taper over the full width. If the taper exceeds the service limit the crankshaft must be renewed.

Standard taper	*0.00008 in (0.002 mm)*
Service limit	*0.0002 in (0.004 mm)*

3   Support the crankshaft at each end on vee-blocks or between centres. Rotate the crankshaft and measure the runout at the centre main bearing journal, using a dial gauge. If the runout exceeds the service limit, the crankshaft must be renewed, or placed in the hands of a specialist for straightening.

Standard runout	*0.0012 in (0.030 mm)*
Service limit	*0.0019 in (0.05 mm)*

Bear in mind that the runout is half the actual reading taken.
4   Main bearing shells must be selected by referring to the code numbers scribed on the adjacent crankshaft webs and by the main bearing cap code numbers or letters which are marked on the crankcase. The coding on the crankcase can be found behind the right-hand cam belt jockey pulley heat shield. The Arabic numerals refer to the respective main bearings. The Roman numerals or capital letters indicate the code number. Cross refer the main bearing journal letters and the cap letters or numbers with the accompanying chart for correct bearing selection.
5   Check the security of the ball bearings which are used to plug the oilways in the crankshaft. They occasionally work loose, causing lubrication problems and subsequent bearing failure. A loose ball can be carefully caulked back into place.

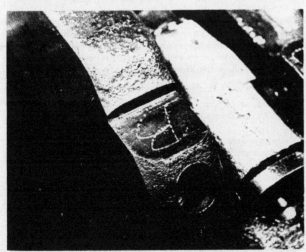

21.6b Big-end journal code letter on web

22.4 Main bearing selection code on crankcase

22.5 Check security of oil way plugs

MAIN BEARING SELECTION

		CRANKSHAFT MAIN JOURNAL SIZE CODES		
CRANKCASE BEARING SUPPORT SIZE CODES	C or 111	BROWN	BLACK	BLUE
	B or 11	GREEN	BROWN	BLACK
	A or 1	YELLOW	GREEN	BROWN
		1	2	3

Fig. 1.9. Main bearing selection chart

23 Connecting rods: examination and renovation

1   It is unlikely that a connecting rod will bend during normal useage, unless an unusual occurrence such as a dropped valve has caused the engine to lock. It is not advisable to straighten a connecting rod; renewal is the only satisfactory solution.

2   The connecting rods do not have small end bearings in the accepted sense. Each gudgeon pin is a high interference fit in the small end eye, the small end bearing surfaces being in the piston bosses.

3   When checking big-end bearing clearances also check the side play of each connecting rod; using a feeler gauge. The correct clearances are as follows:

Standard axial play	*0.0059 - 0.0118 in*
	(0.15 - 0.30 mm)
Wear limit	*0.0158 in*
	(0.40 mm)

If the clearance on any connecting rod exceeds the wear limit that rod must be renewed.

24 Cylinder bores: examination and renovation

1   The usual indication of badly worn cylinder bores and pistons is excessive smoking from the exhausts and piston 'slap', a metallic rattle that occurs when there is little or no load on the engine. If the top of each cylinder bore is examined carefully, it will be found that there is a ridge on the thrust side, the depth of which will vary according to the wear that has taken place. This marks the limit of travel of the top piston ring.

2   As described in Section 17.8 this ridge should be removed from the right-hand cylinder bores before piston removal, if the extent of wear is sufficient to cause damage to the piston rings.

3   Using an internal micrometer, measure each bore for wear. Take measurements at a point just below the upper ridge, at the centre of the bore and about 1 inch from the lower edge of the bore. Take two measurements at each point. If the diameter at any point exceeds the service limit, the cylinders must be rebored and a set of oversized pistons fitted.

Cylinder internal diameter	*2.8346 - 2.8352 in*
	(72.00 - 72.015 mm)
Wear limit	*2.8386 in*
	(72.1 mm)

Again measure each bore and check for taper over the maximum piston travel. Rebore if the taper on any cylinder exceeds the service limit.

Cylinder taper	*0.0003 - 0.0004 in*
	(0.007 - 0.012 mm)
Wear limit	*0.0019 in*
	(0.05 mm)

Check for ovality by measuring in the manner described for checking bore size. If ovality exceeds the wear limit, a rebore is necessary.

Standard ovality	*0.0039 in (0.1 mm)*
Wear limit	*0.0059 in (0.15 mm)*

4   Honda can supply pistons in four oversizes: 0.010 in (0.25 mm); 0.020 in (0.50 mm); 0.030 in (0.75 mm); 0.040 in (1.0 mm).

5   Assuming that the checks detailed above have been carried out and satisfactory results obtained, check each bore visually. Inspect for score marks or other damage that may have resulted from an earlier engine seizure or displaced gudgeon pins. A rebore will be necessary to remove any deep scores.

6   Check the cylinder block upper mating surface for warpage. This can be done using a ground surface plate or bar if one is available, or by using a sheet of plate glass. Do not use ordinary window glass as this has an unground surface. Plate glass is of the type used in old mirrors and as a surface for some table tops. Use a feeler gauge between the cylinder block and surface plate to ascertain warpage.

Standard warpage	*0.0032 in (0.08 mm)*
Maximum warpage	*0.0039 in (0.1 mm)*

If warpage exceeds the service limit the cylinder block surface will require surface grinding. This is a specialised operation. The crankcases should be returned to a Honda service agent for renovation.

7   If the cylinder bores do not require attention but new piston rings are to be fitted (as is usual after engine dismantling), it is advised that each cylinder is 'glazebusted' before reassembly. This will increase the rate at which the new rings bed in, increasing compression, and will also improve cooling. 'Glazebusting' is a specialised operation and should be entrusted to an expert.

8   Inspect the coolant cavities in the water jacket of each cylinder bank. 'Furring up' around the transfer orifices in the mating surfaces can be removed by the careful use of a scraper. Heavy

deposits within the jacket can most easily be removed after the engine has been reassembled and refitted into the frame, using a special flushing fluid or compound added to the coolant. Ensure that the type used is suitable for aluminium castings.

25 Pistons and rings: examination and renovation

1   If a rebore is necessary, ignore this Section where reference is made to piston and ring examination since new components will be fitted.

2   If a rebore is not considered necessary, examine each piston closely. Reject pistons that are scored or badly discoloured as the result of exhaust gases bypassing the rings. Remove each ring either by carefully opening each ring using the thumbs, or by placing three thin strips of tin between the ring being removed and the piston (see illustration). The oil scraper ring comprises a special crimped ring 'sandwiched' between two thin plain rings. Special care must be taken when removing this ring.

3   Remove all carbon from the piston crowns using a blunt instrument which will not damage the surface of the piston. A wood chisel with the cutting edge slightly dulled is a suitable tool. Clean away all carbon deposits from the valve cutaways and finish off with metal polish to produce a smooth shiny surface. Carbon will not adhere so readily to a polished surface.

4   Measure each piston ring groove width, which should be as follows:

Top and second ring	0.0591 - 0.0599 in
	(1.5 - 1.52 mm)
Wear limit	0.0630 in
	(1.6 mm)
Oil ring	0.1104 - 0.1110 in
	(2.805 - 2.82 mm)
Wear limit	0.1142 in
	(2.9 mm)

5   Generally, when an engine is stripped down completely the piston rings are renewed as a matter of course unless the rings have only been fitted for a short time. If ring life is such that renewal is not warranted automatically, the rings should be examined as follows. Check that there is no build up of carbon on the inside surface of the rings or in the ring grooves of the pistons. Any build up should be removed by careful scraping. An old broken ring, the end of which has been ground to a chisel profile, is useful for this. Replace each ring in its respective groove and measure the ring side play using a feeler gauge. The clearances should be as follows:

Top and second ring	0.0008 - 0.0018 in
	(0.02 - 0.045 mm)
Wear limit	0.0059 in
	(0.15 mm)

There is no measureable side clearance on the oil ring as the crimped ring spring loads the plain rings.

6   Place each ring into the cylinder bore separately and measure the end gap. Push the ring down from the top of the bore, using the piston skirt, so that the ring remains square to the bore and is positioned about 1½ inches from the top. If the end gap exceeds the wear limit on any ring, the rings should be renewed as a set.

Top and second	0.0098 - 0.0158 in
	(0.25 - 0.40 mm)
Wear limit	0.0276 in
	(0.7 mm)
Oil ring	0.0079 - 0.0354 in
	(0.2 - 0.9 mm)
Wear limit	0.0433 in
	(1.1 mm)

7   The pistons are of the 'forged' type. This misleading designation indicates that unlike most motorcycle pistons, the bearing surface of each gudgeon pin is in the two piston bosses and not as is usual in the small end eye or bearing. The gudgeon pin is a high interference fit in the small end eye. Removal and replacement requires the use of a special press and guide mandrel. When piston renewal is necessary or if excessive play has developed between the gudgeon pin and piston bosses, the pistons complete with connecting rods should be returned to a Honda repair agent for inspection and replacement. The correct clearance between the gudgeon pin and piston bosses can only be checked accurately after removal of each pin. The clearances are as follows:

Gudgeon pin outside diameter	0.66929 in ± 0.00011 in
	(17.00 ± 0.003 mm)
Piston boss inside diameter	0.6696 in − 0.6697 in
	(17.010 - 17.016 mm)
Gudgeon pin/piston boss max. clearance	0.0008 in
	(0.050 mm)

8   Check the outside diameter of each piston at the skirt by taking a measurement at 90° to the gudgeon pin. If the piston is worn to below the service limit it must be renewed.

Piston skirt outside diameter	2.8325 - 2.8335 in
	(71.945 - 71.97 mm)
Wear limit	2.8288 in
	(71.85 mm)

Place each piston in its cylinder bore and using a feeler gauge, check that the clearance is within the specified limits.

Cylinder/piston clearance	0.0012 - 0.0028 in
	(0.03 - 0.07 mm)
Wear limit	0.0059 in
	(0.15-mm)

9   The piston crowns will show whether the engine has been rebored on some previous occasion. All oversize pistons have the rebore size stamped on the crown. This information is essential when ordering replacement piston rings or in reboring.

10   The letter stamped on each piston ring, indicates the top of the ring. It is essential that each ring be replaced so that it is the correct way up.

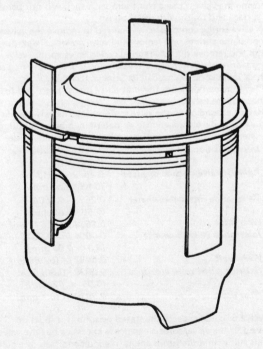

Fig. 1.10. Method of removing and replacing piston rings

25.1 Piston/connecting rod assembly

26 Cylinder head and valves: dismantling, examination and renovation

1   Before examination and renovation of the cylinder heads can be carried out, the camshafts must be removed and the contact breaker assembly detached as follows:

The contact breaker housing is fitted to the rear of the left-hand cylinder head. Remove the two screws which retain the housing and free the low tension cable from the wiring clip. The housing, complete with contact breakers, will pull off the shaft. Loosen and unscrew the Automatic Timing Unit (ATU) bolt from the end of the camshaft and draw the ATU away. Push the radial drive pin from position in the shaft. The camshafts can now be removed, using the same procedure for each.

Remove the six bolts which retain the rocker arm/spindle support and lift the support from position. The support is located on two hollow dowels which should be removed to avoid loss. Removal of the support will free the camshaft, which can be lifted out of the half bearings, together with the oil seals.

Remove the carbon from the combustion chambers before removing the valves. Use a blunt scraper, which will not damage the surface and finish with metal polish.

2   A valve spring compressor is required to remove the valves. Compress the springs and remove the valve collets. Relax the springs and remove the valve spring collar, valve spring, valve spring seats and valve. Keep these components together in a set. Each valve must be replaced in its original location.

3   Clean the carbon from the inlet and exhaust ports and from the head of the valve.

4   Before attending to the valve seats, check the valve guide and stem wear. Valve clearance may be determined by subtracting the valve stem diameter from the internal diameter of the guide.

Inlet valve outside diameter	0.2591 - 0.2595 in (6.58 - 6.59 mm)
Exhaust valve outside diameter	0.2579 - 0.2583 in (6.55 - 6.56 mm)
Valve guide internal diameter	0.2598 - 0.2606 in (6.60 - 6.62 mm)
Wear limit	0.2614 in (6.64 mm)
Inlet guide valve clearance	0.0004 in - 0.0016 in (0.01 - 0.04 mm)
Wear limit	0.0032 in (0.08 mm)
Exhaust guide valve clearance	0.0020 - 0.0028 in (0.05 - 0.07 mm)
Wear limit	0.0039 in (0.1 mm)

If a valve or guide exceeds the stated wear limit it must be renewed. If the valve/guide clearance is excessive but the valve stem is still within the limits and is in good condition, the guide alone may be renewed. The guides may be drifted from position

using a double diameter drift of the correct diameter after the cylinder head has been heated to 80° - 100°C. Heating in an oven is preferable as using a blow torch may distort the heat due to uneven rises in temperature. Replace the new guides using the same drift and with the cylinder head at a similar temperature. A new guide will require reaming in order to bring the valve/guide clearance within the specified tolerances. It is probable also that the valve seat will require cutting in order that the valve seat/guide alignment is exact.

5   Check also that the valve stem is not bent, especially if the engine has been over-revved. If it is bent, the valve must be renewed.

6   The valves must be ground to provide a gas-tight seal, during normal overhaul, or after recutting the seat or renewing the valve.

7   Valve grinding is a simple, but laborious task. Smear grinding paste on the valve seat and attach a suction grinding tool to the valve head. Oil the valve stem. Rotate the valve in both directions, lifting it occasionally and turning it through 90°. Start with coarse paste if the seats are badly pitted and continue with fine paste until there is an unbroken matt grey ring on each seat and valve. After many re-grinds, the valve seat may become pocketed, when it should be re-cut. Wipe off very carefully all traces of grinding paste. If any remains in the engine, it will cause very rapid wear!

8   Valve sealing may be checked by installing the valve and spring and pouring paraffin down the port. Check that none leaks past the valve seat.

9   Check that the valve collets seat well in their grooves. Also check the valve spring collar. Renew any defective part.

10   Check the free length of the valve springs. Renew when below the service limit. Alternatively, compare with new springs.

11   Fit new oil seals to each valve guide after checking that the valve spring seats are in place. Oil the valve stem, and replace the valve. Fit both valve springs with the close coils next to the cylinder head, followed by the valve spring collar. Compress the springs and fit the valve collets. Relax the springs, ensuring that the collets remain in position. Seat the collets firmly with a few blows squarely on the valve stem with a soft hammer. The cylinder head should **not** be resting flat on the bench when doing this.

12   Check the cylinder heads for warpage, using a surface plate or plate glass. If warpage exceeds the service limit the cylinder head will have to be ground or renewed.

Cylinder head warpage	0.0032 in (0.08 mm)
Maximum warpage	0.0039 in (0.1 mm)

Cylinder head/cylinder block mating surface warpage is usually due to uneven tightening of the cylinder head bolts.

26.2a Use compressor to remove valve collets

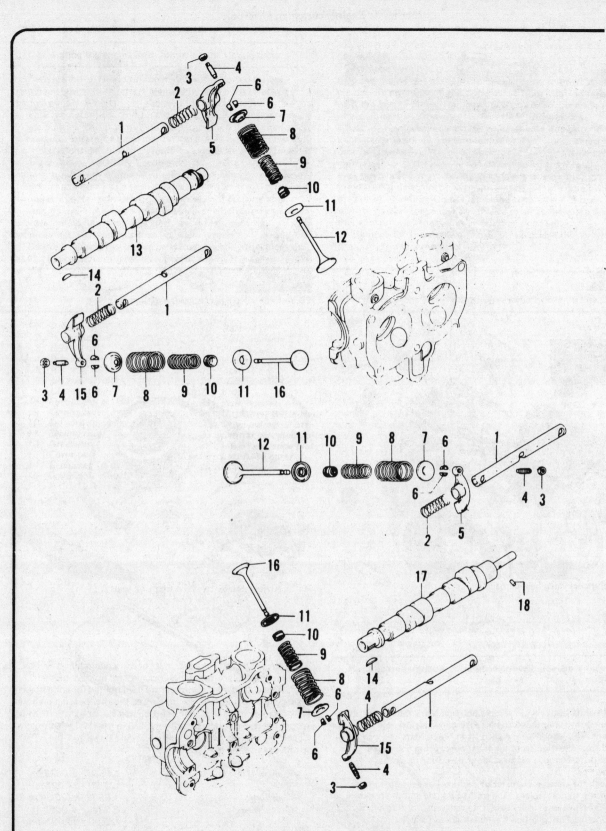

Fig. 1.11. Camshaft and rocker assembly

1	Rocker shaft - 4 off	5	Rocker arm - 4 off
2	Spring - 8 off	6	Valve cotter - 16 off
3	Tappet adjusting nut - 8 off	7	Spring cap - 8 off
4	Tappet adjusting screw - 8 off	8	Outer spring - 8 off

9	Inner spring - 8 off
10	Valve stem seal - 8 off
11	Spring seat - 8 off
12	Inlet valve - 4 off
13	RH camshaft

14	Woodruff key - 2 off
15	Rocker arm - 4 off
16	Exhaust valve - 4 off
17	LH camshaft
18	Drive pin

26.2b ...and inner and outer valve spring

26.3a Inlet valve has concave head ...

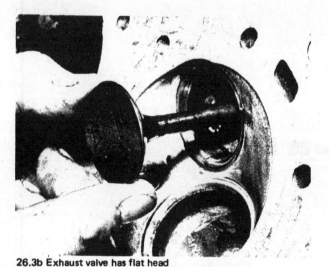

26.3b Exhaust valve has flat head

26.11 Always renew the valve stem oil seals

27 Rocker arms, spindles and camshafts: examination and renovation

1   Dismantle both rocker spindle holders. Mark the rocker arms and spindles so that they can be refitted in their original positions. The spindles are a push fit in the holders and can be tapped from position from the front edge, using a suitable parallel drift. Remove each rocker arm and spring as they become free.

2   Check the outside diameter of each rocker spindle and the bore of the spindle holes in the spindle holder and the rocker arms. Excessive wear will necessitate renewal of the relevant components. The specified clearances are as follows:

Rocker spindle outside diameter	0.5498 - 0.5505 in
	(13.966 - 13.984 mm)
Wear limit	0.5488 in (13.94 mm)
Spindle hole internal diameter	0.5512 - 0.5519 in
	(14.00 - 14.018 mm)
Wear limit	0.5532 in (14.05 mm)

3   In order to check the camshaft bearing diameters it will be necessary to replace the spindle/rocker arm holders on their respective cylinder heads. Insert the two hollow locating dowels into each head and refit the holders so that the bearing oil holes are on the exhaust side. Refit and tighten the holder bolts to the recommended torque which is 84 - 120 in lbs (100 - 140 kg cm). Check the bearing diameters which should be within the specified limits as follows:

Right and left bearing	1.0630 - 1.0638 in
	(27.00 - 27.021 mm)
Wear limit	1.0623 in (26.98 mm)
Centre bearing	0.9843 - 0.9851 in
	(25.00 - 25.021 mm)
Wear limit	0.9835 in (24.98 mm)

If the bearing surfaces are scored or appear to have 'picked-up' due to lubrication failure, it is likely that the bearing surfaces will have to be renewed. Replacement bearings are supplied complete with a cylinder head, rocker spindle holder and valve guides.

4   Measure the journal diameters on each camshaft and check

that they come within the prescribed specifications which are as follows:

Right and left journal outside diameter	1.0612 - 1.0619 in (26.954 - 26.970 mm)
Wear limit	1.0595 in (26.91 mm)
Centre journal outside diameter	0.9824 - 0.9831 in (24.954 - 24.970 mm)
Wear limit	0.9807 in (24.91 mm)

The camshaft journal/bearing clearances can be found by subtracting the measurement taken from each camshaft journal from that of each respective bearing diameter. 'Plastigauge' can also be used to ascertain the clearance. In this case separate measurement of the journals and bearings will not be necessary. The correct journal/bearing clearances are as follows:

Right and left-hand bearing clearance	0.0015 - 0.003 in (0.040 - 0.077 mm)
Centre bearing clearance	0.0019 - 0.0034 in (0.050 - 0.087 mm)

5   Inspect the cam lobes for scoring or uneven wear and chipping of the hardened outer surfaces. If a cam lobe is badly worn or chipped it is probable that the rocker arm foot that runs on that lobe is also damaged and will require renewal. Measure the cam lobes at the highest point of lift (maximum diameter). Camshaft renewal will be necessary if one or more lobes does not come within the specified limits.

Exhaust lobe wear limit	1.437 in (36.5 mm)
Inlet lobe wear limit	1.448 in (36.8 mm)

28 Oil seals and 'O' rings: examination and replacement

1   It is recommended that the oil seals and 'O' rings be renewed whenever the engine is dismantled. This is particularly important with those seals and rings that are non-accessible where major dismantling would be required for subsequent replacement. Seals which have previously given faultless service often begin to weep after reassembly due to damage during handling. This is particularly so with seals which are removed or replaced over splined shafts.
2   In most cases the oil seals are fitted between two separate cases, and can be removed easily after separation of the case. Other seals are a light drive fit in their housings and can be prised out of position using a screwdriver. Removal in this manner will invariably render the seal useless.

29 Cam belt pulleys and cambelts: examination and renovation

1   Wear of the cam belt pulleys is minimal and they will not require renewal until a considerable mileage has been covered. Check the profile of the teeth against that of a new belt. If excessive wear is evident, the pulley wheel should be renewed without question as a worn pulley will accelerate wear of the cam belt and cause inconsistencies in the valve timing.
2   The toothed camshaft belts consist of fibreglass cores binding a synthetic rubber cover. Check wear of the teeth, renewing if there is any doubt about their condition. The belts are immensely strong under extension forces but are easily damaged if mis-handled. Do not bend any part of the belt to a radius of less than approximately 1.0 in (25 mm) and never try and bend the belt about the centre line. Minor scores or scratches caused by removal using screwdrivers or other levers will usually develop into severe damage causing early failure of the belt. Again, renew if there is any cause for doubt.
3   The belt jockey wheel tensioners will rarely give trouble and only require renewal if the outer periphery becomes damaged, thereby causing wear of the pulley belts, or if the centre bearings fail. In either case, the complete wheel and bracket must be

renewed. The extension springs fitted to the pulleys automati-cally control the tension of each belt when they are adjusted manually. If it is evident that the springs have become weak and are not giving sufficient load, they should be renewed.

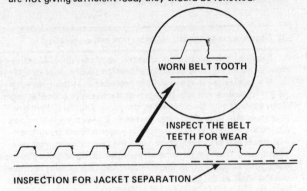

Fig. 1.12. Checking toothed belt wear

30 Primary drive chain: examination and renewal

1   Examine the primary drive chain for wear and loose or broken side plates. The chain is of the 'Hy-vo' type and therefore does not have rollers. There are no specific figures available by which wear can be assessed, but some indication of wear can be obtained by the amount of wear in the chain guides. If the chain guides are very badly worn it is evident that the chain has stretched. Renew both guides and the chain.

31 Clutch assembly: examination and renovation

1   Check the teeth on the duplex sprocket to the rear of the clutch outer drum for chipping or wear. As with the other chain drives in this engine, the components work under almost ideal conditions and will therefore have a very long service life.
2   Carefully clean all the clutch plates. Check the plain plate warpage by placing each plate on a flat surface and measuring with a feeler gauge. If the plates show signs of 'blueing' or bad scoring they should be renewed.

Maximum warpage - plain plate	0.0118 in (0.3 mm)

3   Measure the thickness of each friction plate with a vernier gauge. The correct dimensions are as follows:

Standard thickness	0.1347 - 0.1410 in (3.42 - 3.58 mm)
Wear limit	0.1260 in (3.2 mm)

It is probable that all the friction plates will wear at a similar rate and should therefore be renewed as a complete set. Check that the tongues that locate in the outer drum and clutch centre are not worn; any serious burrs or indentations mean renewal. Very small burrs can be removed with a stone or a fine cut file. **Note:** Do not remove too much metal since the tongues will then be of unequal width and spacings, thus they will not take up the drive evenly and will wear the aluminium alloy clutch housing and centre.
4   Examine the clutch outer drum and centre boss grooves, removing any small burrs with a blind edge filé. Burrs left in the two components will prevent the clutch from disengaging smoothly, causing noisy gear selection.
5   Measure the free length of each clutch spring. After prolonged service, the springs will take a permanent set (shorten) and will therefore exert less pressure.

Spring free length	1.3976 in (35.5 mm)
Service limit	1.3386 in (34.2 mm)

6   Clean and regrease the clutch operating mechanism, which is contained in the clutch outer cover. The mechanism will give little trouble, if it is greased regularly. If necessary, the lifting balls can be renewed if they become pitted or develop flats.

32 Gearbox components: examination and renovation

1   Examine the gearbox components very carefully, looking for signs of wear, chipped or broken teeth and worn dogs or splines.
2   If a gear requires renewal, it is probable that the gear with which it meshes will also be worn or damaged in a similar manner. Both gears should be renewed at the same time to prevent problems due to uneven wear, between a new component and a partially worn one. Removal of the gear pinions from the mainshaft can only take place after the front bearing has been withdrawn. This can be done using a two or three legged sprocket puller positioned so that the 33 tooth top gear pinion is drawn off the shaft at the same time. The remaining gear pinions can then be removed in the same way as those on the layshaft, by removal of the various splined thrust washers and circlips. Carefully note the sequence of gears, washers and circlips to aid replacement. Their correct positioning is essential.
3   If the fit of any of the gears on their shafts is suspect, measure the relative shaft diameters and gear pinion internal diameters and compare them with the following specifications. Renew where necessary.

> *Shaft/mainshaft 4th gear and layshaft 2nd and 3rd gear pinion clearance and;*
> *Bush/mainshaft 4th gear, layshaft 1st gear clearance*
> *0.0016 - 0.0032 in (0.040 - 0.082 mm)*
> *Wear limit 0.0072 in (0.182 mm)*

4   Clean the journal ball bearings thoroughly in petrol and check for wear. Renew bearings if roughness is felt when they are rotated or if pitting or up and down play (radial) is evident.
5   Roll the selector fork rod on a piece of plate glass to check for straightness. If this is satisfactory, measure the outside diameter of the rod and the inside diameter of the selector fork bores. Also check that the finger thickness of the forks is within the recommended limits as follows:

Selector rod outside diameter	*0.5105 - 0.5112 in (12.966 - 12.984 mm)*
Service limit	*0.5079 in (12.90 mm)*
Selector fork internal diameter	*0.5118 - 0.5125 in (13.00 - 13.018 mm)*
Service limit	*0.5134 in (13.04 mm)*
Fork finger thickness	*0.2520 - 0.2559 in (6.4 - 6.5 mm)*
Wear limit	*0.2402 in (6.1 mm)*

6   If the machine tends to jump out of gear it is most probably due to worn dogs on the gear pinions. Difficulty in selecting gears is usually caused by bent selector forks or worn fork guide channels in the change drum.

Groove width	*0.5118 - 0.5125 in (13.00 - 13.018 mm)*
Wear limit	*0.5134 in (13.04 mm)*

Visual indication of wear in the change drum channels is usually most evident at abrupt changes of curvature.
7   Wear in the gear selector mechanism can only be rectified by direct replacement of the parts concerned. This applies equally to those components on the outside of the gearbox, such as the stop arms. If jumping out of gear or over selection has been experienced, renew the stop arm and change the claw springs as a first step in eliminating the causes. This can be done while the engine is still in the frame. If the pins in the end of the change drum become worn they too can be renewed after removal of the

end cap. The cap is held by a counter-sunk screw. Note that one pin is of double diameter and of a shorter length than the other four. Breakage of the main change arm centraliser spring can only be remedied after complete dismantling of the engine. For this reason this spring should be renewed if there is the least doubt as to its condition.

33 Primary drive gears: examination and renovation

1   Examine the condition of the primary driven gear upon which the 'Hy-vo' chain runs. The chain gear operates under almost ideal conditions in that lubrication is supplied in adequate quantities, and the components do not suffer from condensation as is often the case with primary drive systems. Check the primary drive gear on the crankshaft at the same time. Wear in either component will necessitate renewal. Unfortunately, if the drive gear requires replacement, the crankshaft will also have to be renewed as the two components are considered as a single unit and are not supplied separately.
2   Check the journal ball bearing and the two needle roller bearings in the primary driven gear assembly for wear. The ball bearing can be drawn from position using a two legged sprocket puller. The needle roller bearings can be drifted out.

32.2a Pinions are removable from shafts, after detaching circlips

32.2b Layshaft complete

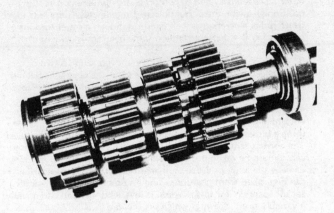

32.2c Mainshaft complete

33.2 Journal ball bearing half-clip grooves are offset

32.2d Oil ways MUST align when refitting inner races

34 Final drive output shaft: examination and renovation

1   Check the shaft double gear for worn, broken or chipped teeth. Renew if necessary. Note that the smaller of the two gears is in fact a splined boss which engages with the moving portion of the shock absorber unit.
2   The shock absorber system comprises two cam faced bosses loaded by a helical spring, which is retained under tension by an end cap and two split cotters. Removal of the spring requires the use of a special tool by which the spring may be compressed and the retaining cotters displaced. This system of spring retention is also used on most types of rear suspension unit. If the special tool is not available, the type of clamp used for suspension unit dismantling could probably be used. Alternatively, a pair of scarf joint clamps as used by carpenter joiners can also be utilised.
3   Compress the main spring sufficiently to release the retaining cotters and then release the tension slowly. Do not overcompress the spring as it may be damaged permanently. Examine the cam faces for excessive indentation or flaking. Although alteration of the cam profile due to wear will have little effect on the performance of the shock absorber it is probably wise to renew the two mating components if wear has progressed to a point when the case hardening has been worn through.
4   Measure the free length of the helical spring, renewing if it fails to meet the specification for length.

Spring free length	*4.3661 in (110.9 mm)*
Service limit	*3.937 in (100 mm)*

35 Starter clutch and alternator damper unit: examination and renovation

1   The starter clutch assembly is housed in the rear of the alternator generator rotor. Check the condition of the three engagement rollers and the boss on the starter driven sprocket with which they engage. If necessary, the complete starter clutch can be removed from the rear of the rotor after unscrewing the three countersunk screws. It is unlikely that any part of the starter clutch will require attention, as the components are only subjected to a limited amount of use.
2   Examine the condition of the starter drive sprocket, the driven sprocket and the drive chain. In the unlikely event of worn components, renewal is the only method of renovation.
3   The alternator rotor and starter clutch are mounted on a double gear shaft incorporating a rubber segment cush drive shock absorber system. The gear pinions and cush drive unit are retained by a short helical spring in a manner similar to that used

33.1 Primary driven gear runs on needle roller bearings

on the final drive output shaft. If care is taken not to damage the aluminium bearing housing, large G-clamps can be used to compress the spring and release the cotters.

4   Examine the condition of the two gears, the damper plate and stopper plate, renewing if required. Check the cush drive rubbers for hardening or compaction and renew if either is evident. Measure the free length of the helical spring and renew if it has taken a permanent set.

Spring free length   *1.023 in (26 mm)*

5   When reassembling the unit, ensure that the gears engage correctly with the damper plate pins. Before reassembly, check the journal ball bearing for wear. Renew the bearing if any up and down (radial) play or roughness and pitting is evident.

36 Journal ball bearings: examination, removal and replacement

1   When the engine is in a completely dismantled state check the condition of all journal ball bearings.
2   Wash each bearing thoroughly with petrol so that all old oil and any foreign matter has been removed. Allow the bearings to dry. Test the bearing for roughness by spinning the outer race. Any roughness or snatching indicates wear or pitting in the races.

The bearing should therefore be renewed. Bearings which are an interference fit in their housings should be tested in situ as the outer races are designed to compress slightly when correctly positioned, giving the correct original tolerances. There should be no radial clearance on ball bearings though a small amount of side play is acceptable and on some bearings is evident even when new.
3   Bearings fitted to shafts may be removed, using a two or three legged sprocket puller. In some cases the clearance behind the bearing is not sufficient to allow purchase of the puller legs. In these cases a special ball bearing puller tool should be used.
4   Removal of bearings within a housing in the crankcase or in detachable housings should only be made after heating the case - using a blowtorch or by placing the casing in question in an oven. The correct temperature is 100° - 150°C. If a blowtorch is used care should be taken not to overheat the case locally as this may damage the alloy or cause permanent distortion. Most bearings can be tapped from position, using a suitable drift. The use of socket spanners for this purpose is invariably decried and equally invariably used. Bearings which are located in blind housings, as is the case with the gearbox layshaft front bearing, can be removed after heating the case to the correct temperature and tapping the casing on a block of wood. The bearing will then fall free.

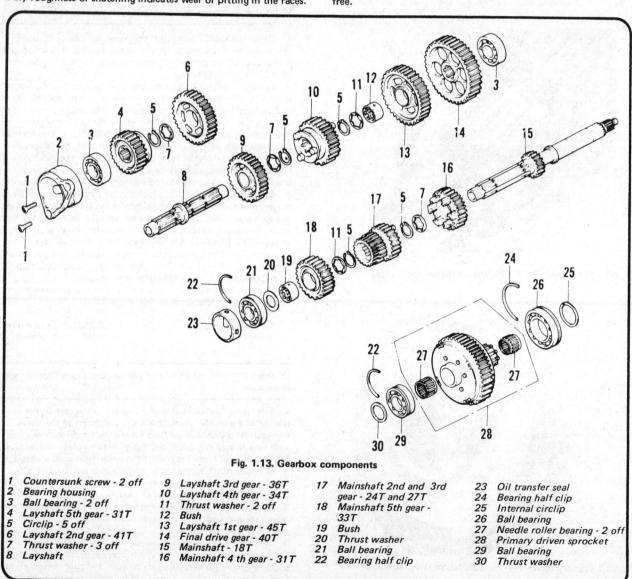

Fig. 1.13. Gearbox components

1	*Countersunk screw - 2 off*	9 *Layshaft 3rd gear - 36T*	17 *Mainshaft 2nd and 3rd*
2	*Bearing housing*	10 *Layshaft 4th gear - 34T*	*gear - 24T and 27T*
3	*Ball bearing - 2 off*	11 *Thrust washer - 2 off*	18 *Mainshaft 5th gear -*
4	*Layshaft 5th gear - 31T*	12 *Bush*	*33T*
5	*Circlip - 5 off*	13 *Layshaft 1st gear - 45T*	19 *Bush*
6	*Layshaft 2nd gear - 41T*	14 *Final drive gear - 40T*	20 *Thrust washer*
7	*Thrust washer - 3 off*	15 *Mainshaft - 18T*	21 *Ball bearing*
8	*Layshaft*	16 *Mainshaft 4th gear - 31T*	22 *Bearing half clip*

23	*Oil transfer seal*
24	*Bearing half clip*
25	*Internal circlip*
26	*Ball bearing*
27	*Needle roller bearing - 2 off*
28	*Primary driven sprocket*
29	*Ball bearing*
30	*Thrust washer*

37 Kickstart mechanism: examination and renovation

1 Owing to the size of the engine and the provision of a perfectly adequate electric start system, it is very unlikely that the kickstart assembly will be used often. Wear will therefore be negligible.

2 The components most likely to fail are the ratchet wheel, which is attached to the end of the crankshaft, the ratchet pawl which is attached to the starter shaft and the return spring.

3 The ratchet wheel is retained on the crankshaft end by a single bolt and is located by two flats ground on the end. Unscrew the bolt to remove the wheel. Dismantle the kickstart shaft assembly and remove it from the casing by removing the guide bolt which is secured by a tab washer and removing the three bolts which retain the spring housing flange plate. Worn parts can now be renewed, as required.

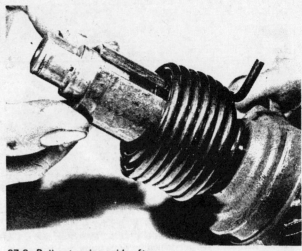

37.3a Pull out spring guide after ...

37.2a Check kickstart ratchet wheel teeth and ...

37.3b ...removing retaining circlip from shaft

37.2b ... the mating ratchet pawl teeth for wear

37.3c Spring will then pull off shaft

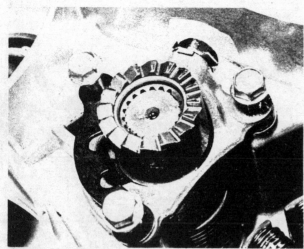

37.3d Kickstart flange retained by three bolts

38 Engine and gearbox reassembly: general

1   Before reassembly is commenced, engine and gearbox components should be thoroughly clean and placed close to the working area.
2   Make sure all traces of old gaskets have been removed and that the mating surfaces are clean and undamaged. One of the best ways to remove old gasket cement, which is needed only on the crankcase and cover joints, is to use a rag soaked in methylated spirit. This acts as a solvent and will ensure the cement is removed without resort to scraping and the consequent risk of damage.
     An alternative method of gasket cement removal is to use a soft wire brush of the type used to clean suede shoes. A considerable amount of scrubbing can take place without any fear of damaging the mating surfaces.
3   Gather all the necessary tools and have available an oil can filled with clean engine oil. Make sure that all new gaskets and oil seals are available; there is nothing more frustrating than having to stop in the middle of a reassembly sequence because a vital gasket or replacement has been overlooked.
     In addition to the tools required for dismantling, a good torque wrench should be acquired. Although in many cases the actual torque figures for a set of bolts may not be critical, it is very important that they are tightened to an equal setting. This will produce a gas or water-tight joint and prevent distortion.
4   Make sure the reassembly area is clean and well lit, with adequate working space. Refer to the torque and clearance settings wherever they are given. Many of the smaller bolts are easily sheared if they are over-tightened. Always use the correct size screwdriver bit for the crosshead screws and NEVER an ordinary screwdriver or punch.

39 Reassembling the engine unit: replacing the gearbox components, pistons and crankshaft

1   Position the right-hand crankcase on the workbench so that it is resting on the cylinder block mating surface.
2   Lubricate the change drum bearing surfaces and introduce it into position, through the hole in the gearbox wall. Insert the layshaft into the gearbox complete with all the gear pinions except the forward most one (top gear, 31T). If the layshaft assembly was dismantled for examination or renewal of gear pinions, it must be reassembled before replacing it in the gearbox. Refer to the accompanying illustration for the relative positions of the gears, washers and circlips. It is very important that the

oil hole in the splined bottom gear pinion bush aligns with the oil hole in the shaft. With the layshaft in position refit the top gear pinion (31T) so that the dogs face inwards. Lubricate the outer journal ball bearing with clean engine oil. Position the bearing, complete with bearing holder, so that the two screw lugs align with the holes in the gearbox wall. Carefully push the bearing holder home. Insert and tighten the two countersunk retaining screws.
3   If the mainshaft was dismantled it must now be reassembled, complete with bearings, before being installed in the casing. As with the layshaft assembly, fit the various gear pinions, washers and circlips by referring to the accompanying illustration. Again the splined bush is provided with an oil orifice which must align with the hole in the shaft. When the bearings have been refitted, check that the exact distance between the outer facing faces of the two journal ball bearings is 6.969 in (177 mm). This is to ensure that the bearing half clips in the crankcase align with the radial locating grooves in the bearing outer races. For the same reason, the bearings must be fitted the correct way round as the radial grooves are offset in the outer race width.
4   Insert the selector fork rod through the gearbox wall, fitting the three selector forks as it is pushed home. It is important that the three selector forks are fitted in the correct order on the shaft and the correct way up. Refer to the accompanying photograph for indication. When the selector fork rod is fully home and the two outer forks are correctly engaged with the pinions on the layshaft, rotate the rod by means of the screwdriver slot in the outer end so that the locating pin holes align. Insert the pin through the crankcase wall and gently tap it home until the head of the pin is just lower than the edge of the crankcase mating surface. Do not knock the pin in further than necessary or subsequent removal will be made difficult.
5   Reposition the crankcase so that access can be made to the right-hand cylinder bores. Refit the piston rings onto the right-hand pistons. When refitting the 'sandwich' type oil scraper ring, the two thin plain rings must be positioned so that their gaps are 0.8 in (20 mm) or more from the corrugated ring gap and more than 1.6 in (40 mm) from each others gaps. The upper rings must be fitted so that the letter mark, which is stamped on one side of each ring, faces upwards. This is important to retain maximum compression. When fitting any piston into its cylinder bore, the rings should be arranged so that the end gaps are approximately 120° apart.
6   Lubricate the cylinder bores with clean engine oil. Insert the pistons complete with connecting rods and rings into cylinders No. 1 and No. 3. A piston ring clamp should be used to compress the rings as they enter the bores. Fitting the pistons without a clamp is possible but the risk of damage to the rings is great as there is no chamfered lead-in. The pistons should be fitted so that the oil hole in each connecting rod faces the top edge of the engine. Note that although oil holes are provided on the connecting rods for No. 1 and No. 3 cylinder, no oil holes are provided in the shell bearings for these rods.
7   Reposition the crankcase so that access to the gearbox can be made again. Lubricate the big-end journals on the crankshaft and fit No. 2 and 4 pistons and connecting rods complete with piston rings. Ensure that the big-end bearing shells are fitted correctly so that the oil holes in the shells align with those in the connecting rods. Fit the connecting rods to the crankshaft so that the oil holes face the top edge of the engine. Tighten the bearing cap bolts evenly to a torque of 24 - 27 ft lb (3.3 - 3.7 kg m), checking as tightening progresses that the bearings remain free.
8   Place the big-end shells on No. 1 and 3 connecting rods in position in the rods and the bearing caps. In the same way fit the main bearing shells into their respective bearing cups. It is absolutely essential that all surfaces of the bearing shells on any part of the engine and the cups into which they fit are perfectly clean. Lubricate all the exposed journals on the crankshaft with clean engine oil. Lubricate and fit the oil seal onto the front end of the crankshaft and position the 'Hy-vo' primary chain on the primary drive sprocket (gear). Lift the complete crankshaft assembly and place it carefully into position in the crankcase.

Insert the main bearing cap hollow locating dowels and fit the main bearing caps so that the arrow on each cap faces the engine top edge. Position the big-end bearings of No. 1 and No. 3 cylinder on the journals and refit the bearing caps. Tighten the main bearing bolts to a torque of 28 - 30 ft lbs (3.8 - 4.2 kg m) and the big-end bearing nuts to 24 - 27 ft lbs (3.3 - 3.7 kg m). Rock the crankshaft occasionally as tightening proceeds to ensure that the bearings do not tie.

9   Insert the gearbox mainshaft bearing half clips in the grooves in the bearing half caps. Lubricate the primary driven gear internal needle roller bearings and fit the gear into the crankcase so that it meshes with the primary drive chain. Insert the mainshaft complete through the primary drive gear and position it correctly so that the bearings locate with the half clips. Install the primary driven gear oil catch plate, lifting the gear as necessary so that the plate can be manoeuvered into position. Fit and tighten the three plate retaining screws. Replace the mainshaft blind end cap so that the small rubber locator tab aligns with the recess in the gearbox wall. This is important as it ensures that the oil passage orifice in the end cap aligns with the feed channel in the gearbox wall.

39.2a Lubricate and insert the gear change drum

39.2b Introduce the layshaft into the gearbox and ...

39.2c ...then fit the 5th gear pinion, dogs facing inwards

39.2d Apply locking fluid to countersunk screws

39.4a Fit selector rod and forks, engaging with gears

39.4b Align radial hole in selector rod and insert pin

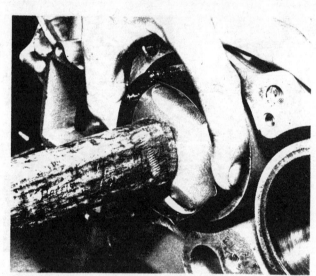

39.6a Fit piston/connecting rod assembly with ...

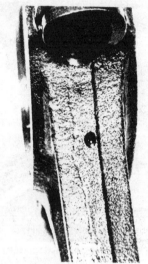

39.6b ... the oil hole facing upwards

39.8a Lower crankshaft into position together with 'HY-VO' chain

39.8b Fit Nos. 1 and 3 big-end bearing caps and ..

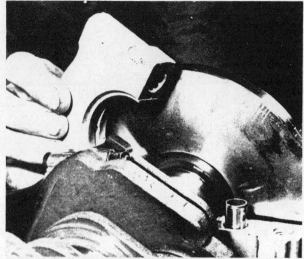

39.8c ... all the mainshaft caps

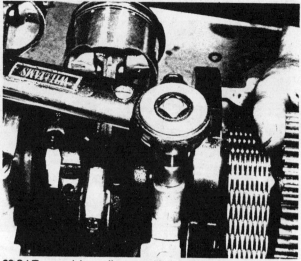

39.8d Torque tighten all nuts and bolts to correct settings

39.9a Fit the primary driven gear to the chain and ...

39.9b ... insert the mainshaft complete

39.9c Slide oil catch plate around chain

39.9d Blind seal locator tab must align with recess

39.9e Gearbox components, general view

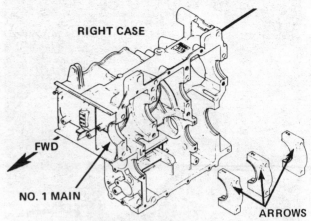

Fig. 1.14. Correct alignment of main bearing caps

40 Reassembling the engine unit: replacing the oil pumps

1   Fit the large oil catch plate into the bottom of the left-hand crankcase and replace the lower most retaining bolt. Fit the primary drive chain guide, noting the distance piece which is fitted below the guide and is retained by the central most bolt. Replace the oil reservoir plate which is retained by two bolts, passing through the crankcase to the rear of the flywheel viewing cap. Coat the two bolts with jointing compound before fitting them.

2   Insert the 8 mm and 14 mm hollow dowels in the main oil pump mounting lug and place a new paper gasket over them. Slide the oil pump driveshaft through the lug until the main pump is correctly positioned. Fit and tighten the pump retaining bolts. Fit a new 'O' ring onto the inner boss of the clutch scavenge pump and position the pump over the shaft end. Push the pump home so that the oil outlet nozzle engages with the orifice in the casing wall. Fit and tighten the three pump retaining bolts. Check free movement of the pumps by rotating the shaft.

3   Grease the gearchange splined shaft and carefully push it through the crankcase wall from the inside. Do not rotate the shaft until it is fully home or the splines will damage the oil seal. Insert the gearchange main change arm into position in the front wall of the gearbox. Ensure that the centraliser spring distance piece is in place and that the two ears of the spring lie one either side of the anchor screw. As the change arm shaft is pushed fully home, fit the shaft ball end so that it engages with the fork on the splined gearchange shaft. Insert the retaining bolt, fitted with the tab washer, so that the bolt end locates with the radial hole in the shaft. Tighten the bolt and bend the tab washer upwards. Tie the main change pawl back against the change arm in a manner similar to that used for dismantling.

41 Reassembling the engine unit: joining the crankcase halves

1   Insert the primary drive change lubrication nozzle into the crankcase wall. Replace the three hollow crankcas locating dowels and the 'O' ring which is fitted to the smallest dowel. Generally, lubricate all the internal engine components, using an oil can
2   Because the pistons on No. 2 and 4 cylinders have to be fitted as the left-hand crankcase is lowered on the right-hand side and because access is obscured, no piston ring clamps can be used. To overcome this the base of each cylinder bore sleeve is heavily chamfered to aid piston replacement. The pistons must be

supported so that they remain parallel and absolutely square as the upper crankcase half is fitted. Wooden blocks can be used for this purpose which are placed between the crankshaft web on one side of each connecting rod and the piston skirt. Alternatively, a forked piece of flat iron can be fabricated (see photograph) to hold one piston while the other piston is supported by hand. Whichever method is chosen, crankcase joining requires two people. So have an assistant at hand.

3   Coat the mating face of the left-hand crankcase half with gasket compound. Honda recommend 'Three Bond' case sealer, but most good quality compounds are satisfactory for this application. Thoroughly lubricate the cylinder bores with engine oil.

4   Lower the left-hand crankcase half down over the pistons until the rings enter the cylinder bores. The piston supports th can then be removed and the casing tapped home, using the palm of the hand. Fit the nineteen crankcase bolts that pass through the right-hand casing. There are three different sizes of bolt: see accompanying diagram for their correct positions. Tighten the bolts down evenly and in as much a diagonal sequence as possible to the following torque settings for each bolt size.

10 mm:	*24 - 27 ft.lb.*	*(3.3 - 3.7 kg.m)*
8 mm:	*216 - 252 in.lb*	*(250 - 290 kg.cm)*
6mm:	*87 - 122 in.lb*	*(100 - 140 kg.cm)*

Invert the crankcase and fit the remaining three crankcase bolts, torquing tightening them to their correct settings, depending on size.

40.1a Refit the oil catch plate ensuring ...

40.1b ... that the spacer collar is not omitted

40.1c Fit the oil reservoir plate

40.2a Refit the main pump first followed ...

40.2b ... by the clutch scavenge pump

40.3a Grease splined change shaft before insertion

40.3b Bend tab washer up after tightening bolt

41.1 'O' ring is fitted to smallest crankcase dowel

41.2a Support one piston securely on a ...

41.2b ... fabricated forked holding tool

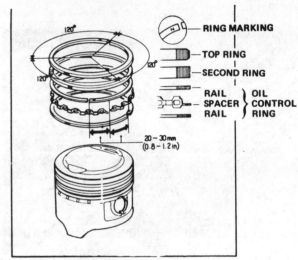

Fig. 1.15. Space the piston rings as shown

41.4 Carefully lower left-hand casing over pistons

42 Reassembling the engine unit: replacing the gear selector mechanism and fitting the front casing

1   Fit the main selector claw over the end of the change drum, with the tongues facing inwards. Make sure that the claw spring 'ears' are on the outside of the outer ears. Fit the centre boss so that the shouldered centres fits into the pivot hole in the claw. Replace the pivot bolt and plain washers. Assemble the two stopper arms, spring and washers in the correct sequence as shown in the illustration and fit the complete assembly onto the pivot stud so that the stopper rollers engage with the end of the change drum. Replace the washers and retaining nut. Untie the main change arm pawl and allow it to engage with the change pins.

2   Temporarily refit the gearchange lever and attempt to select each gear in turn. Rotation of the crankshaft will aid selection. If one or more gears will not engage, check that the external selector mechanism is correctly assembled. If this is found to be correct, the problem will be internal, requiring removal of the left-hand crankcase again. Suspect incorrect fitting of gears and spacer washers if the gear clusters were dismantled for examination.

3   Insert the neutral gear indicator switch into the crankcase, ensuring that the 'O' ring is correctly located in the groove. Refit the retaining tab and bolt.

4   Replace the two hollow dowels into the mating surface of the crankcase onto which fits the engine front cover. Fit new 'O' rings to the oil pump face and insert the feed collar. Replace the three

large 'O' rings and the small 'O' ring and feed collar into their respective housings in the front of the engine.

5 If the water pump has been removed for renewal, it should be replaced before the front cover is replaced. Ensure that the two 'O' rings on the pump body are correctly positioned before fitting the pump into the case. Fit and tighten the three retaining bolts. Place a new front cover gasket in position and refit the cover. Rotate the water pump impeller as the front cover is being fitted, so that the pump shaft engages with the slot in the end of the oil pump shaft. Replace the ten cover screws and tighten them fully in an even and diagonal sequence.

43 Reassembling the engine unit: replacing the alternator drive, starter clutch and alternator

1 Place the engine on the workbench, so that the rear face is upwards. Insert the alternator drive shaft assembly through the crankcase wall and install four of the five retaining bolts. Omit the starter chain guide tab bolt. It is necessary to adjust the backlash between the alternator shaft double gear and the crankshaft to prevent excessive noise when the engine is in service. Adjust the backlash to zero by hooking a spring balance around the alternator shaft and pulling in the direction of the crankshaft to a force of 2.20 lbs. ± 1.10 lbs. (1.0 ± 0.5 kg.) Maintain this force and tighten the four housing bolts to a torque of 87 - 122 in.lb (100 - 140 kg. cm.)

2 Lubricate the starter driven sprocket bearing with clean engine oil and replace the sprocket on the alternator drive shaft so that the boss faces outwards. Mesh the starter drive chain over the sprocket and refit the chain guide tab and retaining bolt. Lubricate the alternator bearings and refit the rotor onto the shaft so that the starter driven sprocket boss enters the starter clutch. Do not tap the starter clutch or alternator in an attempt to force the rollers over the boss. If the rotor is rotated it will slide into place with ease. Fit the rotor retaining bolt, which should be tightened to a torque of 58 - 65 ft.lb. (800 - 900 kg.cm). Mesh the starter motor drive sprocket with the starter drive chain

44 Reassembling the engine unit: replacing the final drive output shaft

1 Insert the output shaft double gear into the crankcase through the crankcase wall. Insert the final drive output shaft through the double gear, so that the shaft engages with the smaller gear (splined boss). Using a new gasket, refit the double gear access hole cover and tighten the four bolts.

42.1a Check spring is correctly fitted and ...

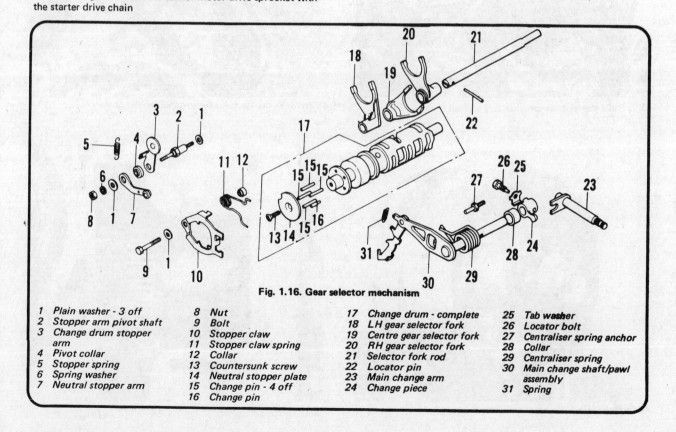

Fig. 1.16. Gear selector mechanism

1	Plain washer - 3 off	8	Nut	17	Change drum - complete	25	Tab washer
2	Stopper arm pivot shaft	9	Bolt	18	LH gear selector fork	26	Locator bolt
3	Change drum stopper arm	10	Stopper claw	19	Centre gear selector fork	27	Centraliser spring anchor
4	Pivot collar	11	Stopper claw spring	20	RH gear selector fork	28	Collar
5	Stopper spring	12	Collar	21	Selector fork rod	29	Centraliser spring
6	Spring washer	13	Countersunk screw	22	Locator pin	30	Main change shaft/pawl assembly
7	Neutral stopper arm	14	Neutral stopper plate	23	Main change arm	31	Spring
		15	Change pin - 4 off	24	Change piece		
		16	Change pin				

42.1b ... replace change claw, tongues facing inwards

42.1c Refit the change drum stop arms

43.2a Refit starter boss sprocket and splined washer

43.2b Mesh the starter drive sprocket to the chain

43.2c Refit alternator rotor and torque centre bolt

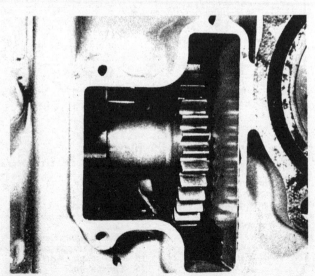
44.1a Replace the final drive double gear and ...

44.1b ... insert the drive shaft to locate the gear

45 Reassembling the engine unit: replacing the clutch assembly

1   Mesh the duplex oil pump drive chain over the sprocket on the clutch outer drum and fit the drum onto the splined clutch shaft. Carefully drift the drum onto the shaft, using a suitable length of wide bore tubing. Before the outer drum is fully home, mesh the oil pump driven sprocket with the chain and fit the sprocket to the pump drive shaft. Replace and tighten the sprocket retaining bolt.

2   Replace the 40 mm internal circlip into the centre of the outer drum, followed by the splined washer which fits over the clutch shaft. Replace the clutch plates one at a time, starting with the clutch pressure plate, a friction plate and then a plain plate. Continue fitting the friction plates and plain plates alternatively. The double thickness damper plate MUST be replaced between the 4th and 5th friction plates, in the place of a normal plain plate. The final friction plate to be fitted is slightly different from the remaining friction plates in that the plate 'ears' are slightly larger. All friction plates must be fitted so that the oil grooves run in a clockwise direction.

3   Replace the clutch centre boss, which may require oscillating slightly before it will enter the clutch plates easily. Refit the curved washer, the tab washer and the special 'peg' nut to the clutch shaft. The curved washer should be fitted so that concave side faces inwards. The outer face is usually marked 'OUTSIDE'. The special nut should be fitted with the chamfered edge inwards. In order that the clutch shaft be prevented from rotating as the special nut is tightened, pressure should be applied to the plates using two or three clutch springs and bolts as described for dismantling. As before, a number of washers should take the place of the clutch lifter plate. Tighten the special nut to a torque setting of 28 - 30 ft. lb. (380 - 420 kg.cm). Do not forget to secure the special nut by means of the tab washer, after tightening.

4   Remove the springs and bolts which were replaced temporarily. Refit all the clutch springs followed by the lifter plate and the spring bolts, which should be tightened to a torque setting of 87 - 121 in.lbs. (100 - 140kg. cm). Insert the clutch lifter piece into the lifter plate bearing.

46 Reassembling the engine unit: replacing the rear engine cover and clutch cover

1   If the kickstart assembly was removed for examination, it must be replaced before the rear cover is fitted. Hook the inner turned end of the kickstart return spring into the hole which passes radially through the kickstart shaft and slide the guide sleeve into

position inside the spring. Fit the guide bolt and a new tab washer to the starter shaft outer holder (flange). Insert the shaft into the rear cover and place the flange onto the shaft so that the guide bolt head faces upwards and to the left. Install the ratchet pawl pressure spring onto the shaft, followed by the pawl, which should be fitted so that the punch mark on the inner recess of the pawl aligns with with the similar punch mark on the starter shaft. Tighten the guide bolt, ensuring that it enters the channel in the ratchet pawl correctly. Bend up the ears of the tab washer against the flats on the bolt head. The ratchet pawl guide plate can now be fitted and the three flange retaining bolts inserted and tightened.

2   Grease the lip of the oil seal through which passes the engine final drive output shaft and place a new gasket on the engine rear cover. Refit the engine rear cover very carefully so that the splined end of the ouput shaft does not damage the oil seal lip . Replace the eleven cover bolts, which should be tightened evenly and in a diagonal sequence. Note that a wiring clip is fitted to the cover bolt located just above the right-hand engine mounting bolt lug. Fit the clutch cover and the retaining bolts. Again, as wiring clip is retained on the cover by the lower of the two nut and studs.

3   Replace the external components of the kickstart mechanism onto their various shafts. The crank and main lever are retained by circlips.

47 Reassembling the engine unit: replacing the cylinder heads and valve gear

1   Each cylinder head unit, complete with valve gear, can be replaced using identical procedure. Commence by replacing either the right-hand or left-hand cylinder head, as chosen.

2   Ensure that the mating surfaces of the cylinder head and block are absolutely clean. Replace the two hollow locating dowels and the oil feed nozzle, which should be fitted with the shorter and narrower end towards the cylinder block. It is important that the 'O' rings are correctly fitted and in good condition. Apply a liquid gasket compound to the cylinder head and cylinder block mating surfaces.Honda recommend' Three Bond No. 4 sealer' for this application. Fit a new cylinder head gasket over the locating dowels and replace the cylinder head. Inster the seven cylinder head retaining bolts. Tighten the six 10mm bolts evenly, in a diagonal sequence, starting from ether of the two most central bolts. Finally torque tighten the bolts down to 38 - 4l ft.lb. (530 - 570 kg.cm). Tighten the seventh bolt (which is 6 mm) to a torque setting of 7 - 10 ft. lb. (100 - 140 kg.cm).

3   Before fitting the camshafts, rotate the crankshaft by means of the alternator rotor bolt so that No. 1 cylinder is at TDC on the compression stroke. Now rotate the crankshaft forwards or backwards through 90°. This will prevent the pistons hitting the valves if the camshafts are so fitted that the valves are open.

4   The camshaft fitted with the tachometer drive gear must be replaced on the right-hand cylinder head. Lubricate the camshaft liberally with engine oil and fit the two oil seals to either end. The spring side of each seal must face inwards. Place the camshaft in position on the cylinder head and fit the rocker arm holder complete with arms etc. Fit the six retaining bolts and tighten them evenly, in a sequence similar to that used for the cylinder heads, to a torque setting of 18 - 21 ft. lb. (250 - 290 kg. cm). Esnure that each oil seal is pushed inwards as far as possible, as the rocker holder is tightened down. To guarantee an oil tight seal between the rocker holder and cylinder head, a small amount of gasket compound should be spread on each face before bolting up. Excess sealer should then be removed from around the joint. Repeat for the second camshaft in the left-hand cylinder head. Replace the heat shields wich fit on the front of each cylinder head and are each retained by two bolts.

48 Reassembling the engine unit: replacing the valve drive assembly and timing the valves

1   Fit the camshaft belt drive pulleys to the forward end of the crankshaft and torque tighten the retaining bolt to 24 - 27 ft.lb

(330 - 370 kg. cm). Replace the woodruff key into the keyway in the end of each camshaft and fit the driven pulleys. The left-hand pulley should be replaced with the boss facing towards the engine and the right-hand pulley boss should face outwards. When tightening the centre bolts, do not allow the pulleys to rotate or the opening valves may come into contact with the pistons. The pulleys may be held by placing a tyre lever through the pulley spokes so that it abuts against one of the heat shield bolt heads.

2　Rotate the cam pulleys so that the arrows align with the index marks on the heat shields. Using the alternator rotor bolt, rotate the crankshaft so that the 'T - 1' mark on the crankshaft flywheel aligns with the index marks on the observation hole.　A length of wire can be placed across the aperture to aid accurate sighting. Without rotating the driven or drive pulleys fit the two timing belts. Old belts must be replaced in the same position as they were on dismantling. Do not strain the belts or prise them into position as damage may result, leading to short belt life.

3　Ensure that the tensioner pulley bolts are loose. Viewing from the front, apply anticlockwise pressure to the left-hand camshaft pulley so that the belt will be slack on the tensioner side. The spring on the tensioner pulley will automatically tension the belt. Without relaxing pressure tighten the tension pulley bolts. Repeat the procedure on the right-hand pulley and then recheck the timing. Replace the timing belt covers, noting that the long screw holds the outer portion of the left-hand cover.

45.1a Replace clutch drum together with pump drive gear

45.1b Do not omit special splined washer

45.2 Locate tab washer when tightening special nut

45.3 Fit lifter plate and tighten bolts fully

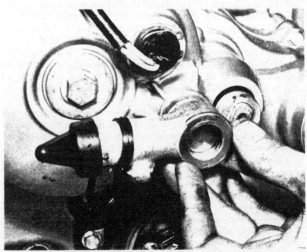

46.3 Grease kickstart assembly shaft before fitting

47.2a Oil feed nozzle must be fitted as shown

47.2b Place cylinder head onto new gasket

47.4a Replace the camshafts and oil seals and ...

47.4b ... refit the rockers and assemblies

47.4c Replace cover extension/heat shields using new gaskets

48.1a Ensure belt guide plates are fitted correctly

48.1b Fit pulleys and tighten centre bolts

48.2a Line up pulleys with arrows facing outwards

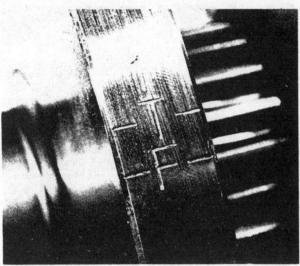
48.2b Timing and firing marks for No. 1 cylinder on flywheel

48.2c Use length of wire to aid accurate sighting

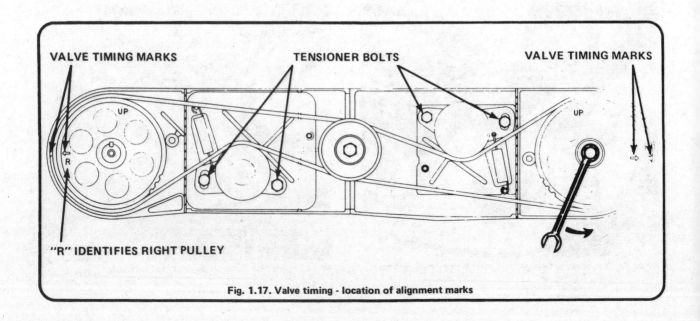

VALVE TIMING MARKS

TENSIONER BOLTS

VALVE TIMING MARKS

UP

UP

R

R

"R" IDENTIFIES RIGHT PULLEY

Fig. 1.17. Valve timing - location of alignment marks

49 Reassembling the engine unit: adjusting the valve clearances and replacing the rocker covers

1   Rotate the engine until both valves on No. 1 cylinder are fully closed and the 'T - 1' mark on the flywheel is aligned with the index marks. With the crankshaft in this position, No. 1 piston is at TDC. Check the following valves by placing a 0.004 in (0.1 mm) feeler gauge between the valve stem head and the rocker adjuster screw

> *No.1   inlet and exhaust*
> *No.3   exhaust*
> *No.4   inlet*

If the gap on any valve is incorrect, loosen the locknut on the adjuster screw and screw the adjuster in or out, as necessary. When adjustment is correct, prevent the screw rotating by using a screwdriver and tighten the locknut. Re-check the settings. Rotate the engine through 360⁰ until the 'T - 1' mark is again aligned with the index marks. No. 2 piston is now at TDC on the compression stroke. Check, and adjust where necessary, the valve clearances on the following valves.

> *No.2   inlet and exhaust*
> *No.3   inlet*
> *No.4   exhaust*

2   Check that all the adjuster screw locknuts have been tightened fully. The recommended torque setting is 9 - 12 ft.lbs. (120 - 160 kg.cm). Refit both rocker covers making sure that the rubber seals are correctly positioned.

50 Reassembling the engine unit: replacing the fuel pump and contact breaker assemblies

1   Fit the fuel pump to the mounting casting and tighten the two bolts. Place the complete assembly in position on the rear of the cylinder head. If difficulty is encountered, rotate the engine so that the pump cam clears the pump arm and the tachometer drive gear meshes easily with the driven gear. Fit and tighten the two retaining bolts.
2   Place the ATU (Automatic Timing Unit) onto the end of the left-hand camshaft, ensuring that the drive pin locates with the slot in the ATU boss. Fit and tighten the centre bolt. Place the contact breaker housing complete with contact breaker assemblies over the ATU and replace the two retaining screws. The housing must be fitted so that the low tension lead is downwards. Secure the lead with the wiring clip on the cylinder head.
3   The points should be adjusted and the ignition timing checked as a matter of course as described in Chapter 4 Section 7. Replace the contact breaker cover.

51 Reassembling the engine unit: replacing the thermostat assembly

1   Fit new 'O' rings to the thermostat housing bypass pipes. Lubricate the 'O' rings with soapy water and fit the pipes to the thermostat housing and manifolds. Insert the hollow locating dowel into the top of the crankcase and fit the 'O' ring. Replace the complete assembly onto the crankcase, using new gaskets on the two pipe manifolds. Insert the six retaining screws. Push the pipes firmly into place and at the same time tighten the screws.
2   Insert the thermostat into the main housing. Replace the thermostat cover, which should be fitted with a new 'O' ring, and tighten the two screws.
3   Apply a waterproof gasket compound to the threads of the electric fan and water temperature switches and replace them in their housing.

49.1a Adjust valves and check gap with feeler gauge

49.2 Replace right-hand cover first

50.1 Fit fuel pump onto drive worm, with care

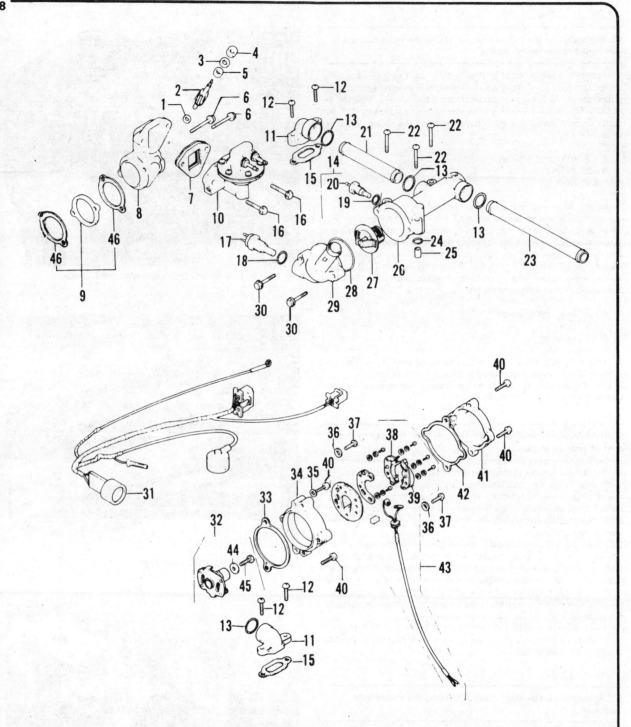

Fig. 1.18. Fuel pump thermostat and contact breaker assembly

1 Thrust collar	13 'O' ring - 4 off	24 'O' ring	36 Plain washer - 2 off
2 Gear shaft	14 Water temperature sender	25 Feed collar	37 Screw - 2 off
3 Washer	switch	26 Thermostat housing	38 RH contact breaker
4 Oil seal	15 Gasket - 2 off	27 Thermostat	assembly
5 Thrust collar	16 Bolt - 2 off	28 Cover 'O' ring	39 LH contact breaker
6 Bolt - 2 off	17 Fan switch	29 Housing manifold	assembly
7 Insulator	18 'O' ring	30 Bolt - 2 off	40 Screw - 4 off
8 Pump casting	19 'O' ring	31 Engine sub harness	41 Points cover
9 Insulator	20 Switch	32 Automatic timing unit	42 Gasket
10 Pump assembly	21 RH transfer pipe	33 Packing piece	43 Points assembly - complete
11 Water manifold - 2 off	22 Screw - 2 off	34 Contact breaker housing	44 Plain washer
12 Screw - 4 off	23 LH transfer pipe	35 Washer	45 Bolt
			46 Gasket - 2 off

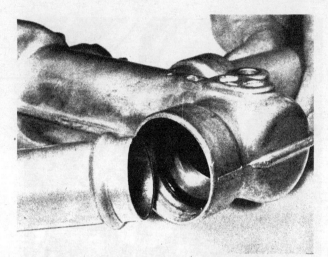

51.1a Apply soapy water to the transfer pipe 'O' rings

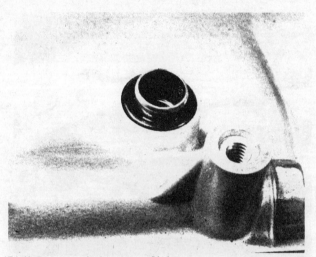

51.1b Do not omit dowel and 'O' ring

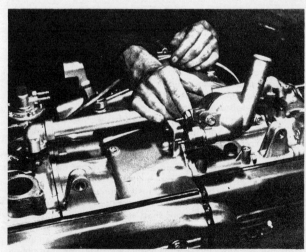

51.1c Fit new gaskets to the pipe manifolds

52 Reassembling the engine unit: replacing the carburettors

1   The carburettors can only be replaced after they have been reassembled into a single unit, the four instruments being attached to the central air box.
2   Place a new 'O' ring in each of the induction manifolds. Used 'O' rings should never be employed as they are invariably stretched and may be displaced during bolting up. Place the complete carburettor assembly in position on the top of the engine, locating each manifold flange with the two studs. Fit the domed retaining nuts. Reconnect the fuel feed line to the front union on the fuel pump.

52.1 Replace carburettors as a complete unit

53 Reassembling the engine unit: replacing the ancillary components

1   Refit the starter motor. Lubricate the 'O' ring around the motor boss with soapy water to aid insertion into the case. The splined starter motor shaft must engage with the starter drive sprocket as it is replaced. Fit and tighten the two screws.
2   Replace the breather box on the rear of the engine. The box is retained by two bolts. Reconnect the hose on the left-hand breather union with the union on top of the crankcase. Do not omit the hose retaining spring clips.
3   Screw the oil pressure warning switch into the top of the crankcase, after applying gasket compound to the threads.
4   Insert the gauze oil trap into the crankcase, making certain that the union locates with the pick-up point in the crankcase. Replace the oil trap cover so that the two pegs on the oil trap locate with the recesses in the cover. Ensure that the cover seal is in good condition before fitting the cover.
5   Place the final drive shaft gaiter in position over the rear of the output shaft.

54 Fitting the engine/gearbox unit into the frame

1   As with engine removal, the use of a trolley jack and the help of two assistants is indispensable for efficient and safe engine replacement. The same system should be adopted, by placing the engine on the jack and wheeling it into position from the left-hand side. Protect the engine sump by inserting a piece of wood between the engine and the jack. During engine replacement a certain advantage is gained in that the engine can be placed on the jack in a

53.1 Mesh starter motor shaft with the drive sprocket

53.2 Replace the breather box and reconnect the pipe

53.3 Apply sealing compound to oil pressure switch threads

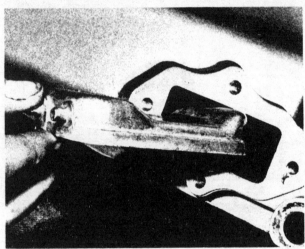

53.4a Insert the gauze oil trap and ...

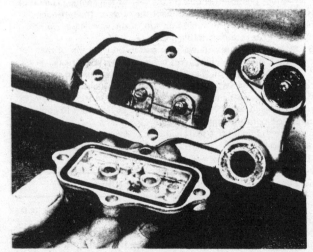

53.4b ... fit the sealing ring and cover

position of optimum balance. This was not possible during removal as the exact position was a matter of guesswork.

2   Move the engine towards the frame until it is close enough to reconnect the following controls: the neutral gear warning switch lead must be reconnected now as it is virtually impossible to do so when the engine is fully installed. The lead terminal is retained by a cross-head screw. Insert the clutch cable through the right-hand side of the casing and reconnect the inner cable with the lifting arm. Clutch adjustment should be carried out now rather than when the engine is fully installed, as access is considerably easier. Loosen the locknuts on the cable upper adjuster screw and lower adjuster screw and screw both adjusters in as far as possible. Loosen the locknut on the clutch operating shaft and, using a screwdriver, turn the shaft in a clockwise direction until it becomes hard to turn (resistance is felt). Rotate the shaft back about ¾ turn and tighten the locknut. Unscrew the cable lower adjuster until play of 0.2 - 0.6 in (5 - 15 mm) can be felt, measured at the ball end of the handlebar lever. Tighten the locknuts. Replace the clutch adjustment cover and tighten the two retaining screws.

3 Reconnect the throttle cables to the throttle pulley. If necessary, move the engine further inwards altering the height of the engine by means of the jack. The throttle pulley will have to be rotated by hand to reconnect the cables easily.

4 Move the engine inwards until it is approximately in the correct final position. Again adjust the engine height as necessary, to avoid fouling the frame. The air box will have to be raised slightly so that the carburettors do not snag. Move the engine forwards slightly so that the final drive shaft universal joint can be placed in line with the output shaft. Move the engine backwards until the output shaft begins to enter the joint. Push the joint forwards fully and refit the retaining circlip.

5 Refit the subframe to the frame left-hand cradle tubes. Replace and tighten the four mounting nuts. Install and lightly tighten the engine mounting bolts in the following sequence, taking care not to damage the bolt threads as they are fitted. Adjust the engine height as necessary so that all lugs and bolt align correctly.

1 Rear left and right lower bolts
2 Front lower cross bolt
3 Rear left and right engine plate frame nuts
4 Rear left and right engine plate engine lug bolts
5 Lower shroud bolts
6 Upper shroud bolts

Note that the main earth lead is fitted between the left-hand rear engine plate and the engine mounting lug, before fitting the engine plate and the engine mounting lug. Before fitting the engine shroud (fan shield), replace the sub-harness which comprises the sensor switch leads. The lead to the oil pressure warning switch must pass below the right-hand thermostat by-pass pipe. Reconnect the remaining push-connections and secure the leads to the rear of the fan shroud by means of the built-in clips. Tighten the engine bolts fully in the sequence given for installation.

6 Reconnect the main wiring leads at the block connectors and connect the starter motor lead to the terminal. Check that all the controls and wires are tracked correctly and will not become chafed or damaged. Refer to the accompanying illustration for recommended positions. Reconnect the contact breaker leads at their snap connectors and fit the H/T leads to the sparking plugs.

7 Loosen and remove the inner bolt from the fuel pump casing flange. Insert the tachometer drive cable, ensuring that the cable end locates correctly and refit the bolt. Attach the fuel pipe to the fuel pump union and tighten the screw clip on the hose.

8 Insert a new exhaust gasket into each exhaust port. If necessary, the gaskets can be distorted very slightly so that they become oval and will remain in the ports. Fit the exhaust pipes into the unions in the silencer. Two people will be required to fit the exhaust pipes into the ports, so that they are fitted simultaneously to prevent tying on the retaining studs. Fit the exhaust pipe flange nuts and tighten them evenly, in a diagonal sequence, until about three threads can be seen below each nut. Take care to tighten these nuts evenly to avoid distortion of the pipe flanges. Ensure that the exhaust/silencer unions are correctly fitted and tighten the clamp socket screws. Tighten the silencer mounting bolts.

9 Turn the front fork onto full lock. Lift the complete radiator assembly into position so that it is supported by the mounting studs. Fit and tighten the retaining dome nuts. Take care when positioning the radiator not to damage the fins. Reconnect the radiator top hose with the thermostat housing manifold and tighten the hose clip. Fit the bottom hose manifold to the hose and then replace the manifold on the front of the water pump housing. Use a new gasket to ensure good sealing. Reconnect the radiator fan leads to the main harness at the 'block' connector, and secure the lead by means of the clip provided. Reconnect the radiator to the reservoir tank.

10 Bolt the air filter box to the carburettor box. Take care not to drop the two bolts into the carburettor box. Replace the air filter and fit the air box cover, noting that it is non-reversible and can only be fitted with the arrow mark facing forwards. Connect the breather pipe from the air filter box to the breather box attached to the crankcase.

11 Refit the gearchange lever onto the shaft.

54.2a Ease the engine in towards the machine

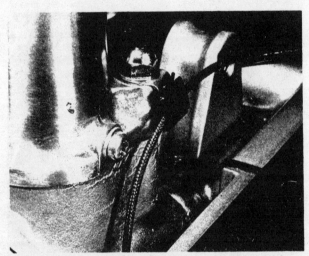

54.2b Reconnect and secure the neutral warning switch lead

54.3 Reconnect throttle and choke cables

54.4a Line up the universal joint and output shaft ...

54.4b ... Move engine backwards and fit the retaining circlip

54.5 Note main earth lead fitted to L/H mounting plate

54.6 Reconnect main starter motor lead

54.7 Insert the tachometer drive cable and tighten bolt

54.8 Always use new exhaust port gaskets

54.10 Reconnect the breather hose to the breather box

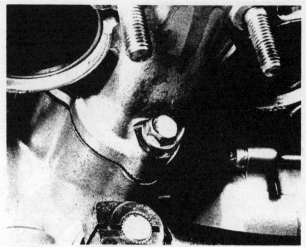

54.11 Replace the gear change lever

55 Starting and running the rebuilt engine

1   Fit a new oil filter into the filter housing and replace the housing on the front of the engine. Note the housing should be fitted so that the aligning marks align with similar marks on the engine front cover. Tighten the housing bolt fully to a torque setting of 20 - 24 ft. lb. (270 - 330 kg. cm). Replace and tighten the engine oil drain plug and the coolant drain plug.

2   Refill the crankcase through the filler orifice to the rear of the fuel pump with the correct specification of engine oil. The engine will accept about 3.7/3.1 US/Imp. qts. (3.5 lites) after complete dismantling and reassebly. Check the level through the sight window in the crankcase.

3   Fill the radiator with the recommended coolant. The capacity after complete dismantling is about 3.0/2.5 US/Imp. qt. (2.8 litres). Replenish through the radiator filler orifice until the level is at the lower end of the filler neck. The cooling system will need to be bled so that all air is remove. This is accomplished by starting the engine, which should be allowed to run at about 900 rpm for 10 minutes, revving up the engine for the last 30 seconds to accelerate bleeding. The coolant level will fall, indicating the expulsion of air. Top up the radiator and replace the cap. If necessary, replenish the reservoir so that the level is just below the upper level mark. Start the engine, and keep it running at a low speed for a few minutes to allow oil pressure to build up and the oil to circulate. If the oil pressure warning lamp is not extinguished, stop the engine immediately and investigate the lack of pressure

4   The engine may tend to smoke through the exhausts initially, due to the amount of oil used when assembling the components. The excess of oil should gradually burn away as the engine settles down.

5   Check the exterior of the machine for oil leaks or blowing gaskets. Make sure that each gear engages correctly, and that all the controls function effectively, particularly the brakes. This is an essential last check before taking the machine on the road.

56 Taking the rebuilt machine on the road

1   Any rebuilt machine will need time to settle down, even if parts have been replaced in their original order. For this reason it is highly advisable to treat the machine gently for the first few miles to ensure oil has circulated throughout the lubrication system and that any new parts fitted have begun to bed down.

2   Even greater care is necessary if the engine has been rebored or if a new crankshaft has been fitted. In the case of a rebore, the engine will have to be run-in again, as if the machine were new. This means greater use of the gearbox and a restraining hand on the throttle until at least 500 miles have been covered. There is no point in keeping to any set speed limit; the main requirement is to keep a light loading on the engine, and to gradually work up performance until the 500 mile mark is reached. These recommendations can be lessened to an extent when only a new crankshaft is fitted. Experience is the best guide since it is easy to tell when an engine is running freely.

3   If at any time a lubrication failure is suspected, stop the engine immediately and investigate the cause. If an engine is run without oil, even for a short period, irreparable engine damage is inevitable.

4   When the engine has cooled down completely after the initial run, re-check the various settings, especially the valve clearances. During the run most of the engine components will have settled into their normal working locations.

55.1a Fit a new oil filter and the housing

55.1b Replenish with the correct grade of oil

57 Fault diagnosis: engine

Symptom	Cause	Remedy
Engine will not start	Defective spark plugs	Remove the plugs and lay on cylinder heads Check whether spark occurs when ignition is switched on and engine rotated
	Dirty or closed contact breaker points	Check condition of points and whether gap is correct
	Faulty or disconnected condenser	Check whether points arc when separated Renew condenser if evidence of arcing
	Valve stuck in guide	Free and clean both stem and valve guide Renew, if binding
	Faulty valve timing	Check and re-set
Engine runs unevenly	Ignition and/or fuel system fault	Check each system independently, as though engine will not start
	Blowing cylinder head gasket	Leak should be evident from oil leakage where gas escapes
	Incorrect ignition timing	Check accuracy and if necessary re-set
	Incorrect tappet clearance	Check and adjust
Lack of power	Fault in fuel system or incorrect ignition timing	See above
	Valve sticking	See above
	Valve seats pitted	Grind in valves
	Faulty piston springs	Renew as a set
	Faulty piston ring	Renew as a set
Engine overheats	Heavy carbon deposit	Decoke engine
	Lean fuel mixture	Adjust carburettors
	Retarded ignition timing	Check and re-set
	See also Chapter 2 Fault diagnosis	
Heavy oil consumption	Cylinder block in need of rebore	Check for bore wear, rebore and fit oversize pistons if required
	Damaged oil seals	Check engine for oil leaks
	Excessive oil pressure	Check pressure relief valve action

58 Fault diagnosis: clutch

Symptom	Cause	Remedy
Engine speed increases as shown by tachometer but machine does not respond	Clutch slip	Check clutch adjustment for free play at handlebar lever. Check thickness of inserted plates
Difficulty in engaging gears. Gear changes jerky and machine creeps forward when clutch is withdrawn. Difficulty in selecting neutral	Clutch drag	Check clutch adjustment for too much free play. Check clutch drums for indentations in slots and clutch plates for burrs on tongues. Dress with file if damage not too great
Clutch operation stiff	Damaged, trapped or frayed control cable	Check cable and replace if necessary. Make sure cable is lubricated and has no sharp bends
	Bent operating pushrod	Check the pushrod for trueness

59 Fault diagnosis: gearbox

Symptom	Cause	Remedy
Difficulty in engaging gears	Selector forks bent Gear clusters not assembled correctly	Replace Check gear cluster arrangement and position of thrust washers
Machine jumps out of gear	Worn dogs on ends of gear pinions Stopper arms not seating correctly	Replace worn pinions Remove right-hand crankcase cover and check stopper arm action.
Gear change lever does not return to original position	Broken return spring	Replace spring
Kickstart does not return when engine is turned over or started	Broken or poorly tensioned return spring	Replace spring or re-tension
Kickstart slips	Ratchet assembly worn	Part crankcase and replace all worn parts

Chapter 2 Cooling system

Contents

Specifications

Cooling system capacity	3.0/2.5 US/Imp qt. (2.8 litres)
Reservoir capacity	0.4/0.35 US/Imp qt. (0.4 litres)
Total capacity	3.4/2.8 US/Imp. qt. (3.2 litres)

Coolant specification
Distilled water 50% mix with ethylene glycol antifreeze (**Warning:** Alchohol based antifreeze must not be used).
Boiling point (50/50 mixture):

Unpressurized	107.7°C (226°F)
Pressurized	125.6°C (258°F)
Pressure cap release pressure	10.7 14.9 psi (0.75 = 1.05 kg/cm^2)

Thermostat

Type	Wax pellet
Begins opening	80° - 84°C (176° - 183°F)
Fully open	93° - 96°C (199° - 205°F)
Valve lift	Minimum of 0.315 in (8 mm) at 95°C (203°F)

Fan switch

Fan on	98°- 102°C (208° - 216°F)
Fan off	93° - 97°C (199° - 207°F)

1 General description

The Honda Gold Wing is provided with a car type cooling system which utilises a water/antifreeze coolant to carry away excess energy produced in the form of heat. The cylinders are surrounded by a water jacket from which the heated coolant is circulated by thermo-syphonic action in conjunction with a water pump drivn off the front of the oil pump drive shaft. The hot coolant passes upwards through a thermostat housing, to the top of the radiator which is mounted on the frame downtubes to take advantage of maximum air flow. The coolant then passes downwards, through the radiator core, where it is cooled by the passing air, and then to the water pump and engine where the cycle is repeated. An electric fan is mounted behind the radiator to aid cooling, when operating conditions demand. The electric fan motor is activated automatically by means of a sensor switch fitted to the thermostat housing. A wax pellet type thermostat is fitted in the system to prevent the flow of coolant through the radiator when the engine is cold, thereby accelerating the speed at which the engine reaches normal working temperature.

The complete system is sealed and pressurised; the pressure being controlled by a valve contained in the spring loaded radiator cap. By pressurising the coolant to approximately 13 psi, the boiling point is raised, preventing premature boiling in adverse conditions. The overflow pipe from the radiator is connected to a reservoir into which excess coolant is discharged by pressure. The expelled coolant automatically returns to the radiator, to provide the correct level when the engine cools again.

2 Cooling system: draining

1 Place the machine on the centre stand so that it rests on level ground. If the engine is cold, remove the radiator cap in the normal manner by pressing the cap downwards and rotating it in an anticlockwise direction. If the engine is hot having just been run, place a thick rag over the cap and turn it slightly until all the pressure has been allowed to disperse. A rag must be used to prevent escaping steam from causing scalds to the hand. If the cap were to be removed suddenly, the drop in pressure could allow the water to boil violently and be expelled under pressure from the filler neck. Apart

from burning the skin the water/antifreeze mixture will damage paintwork.

2   Place a receptacle below the front of the engine into which the coolant can be drained. The container must be of a capacity greater than the volume of coolant which is 3.0/2.5 US/Imp. qt. (2.8 litres). The recommended interval at which the coolant should be renewed is 24,000 miles or every two years. If the coolant is to be re-used, ensure that the container is perfectly clean. Remove the cap from the reservoir and detach the reservoir from position inside the dummy fuel tank. Remove the drain plug from the water pump casing and allow the coolant to drain completely.

3   Cooling system: flushing

1   After extended service the cooling system will slowly lose efficiency, due to the build up of scale, deposits from the water and other foreign matter which will adhere to the internal surfaces of the radiator and water channels. This will be particularly so if distilled water has not been used at all times. Removal of the deposits can be carried out easily, using a suitable flushing agent in the following manner.

2   After allowing the cooling system to drain, replace the drain plug and refill the system with clean water and a quantity of flushing agent. Any proprietary flushing agent in either liquid or dry form may be used, providing that it is recommended for use with aluminium engines. NEVER use a compound suitable for iron engines as it will react violently with the aluminium alloy. The manufacturer of flushing agent will give instructions as to the quantity to be used.

3   Run the engine for ten minutes at operating temperatures and drain the system. Repeat the procedure TWICE and then again using only clean cold water. Finally, refill the system as described in the next section.

4   Cooling system: filling

1   Before filling the system, always check that the drain plug has been fitted and tightened and that the hose clips are tight.

2   Fill the system slowly to reduce the amount of air which will be trapped in the water jacket. When the coolant level is up to the lower edge of the radiator filler neck, run the engine for about 10 minutes at 900 r.p.m., reving up for the last 30 seconds to accelerate the rate at which any trapped air is expelled. Stop the engine and replenish the coolant level again to the bottom of the filler neck. Refill the reservoir up to the 'Full' level mark. Replace the radiator cap, ensuring that it is turned clockwise as far as possible.

3   Ideally, distilled water should be used as a basis for the coolant. If this is not readily available rain water, caught in a non-metallic receptacle is an adequate substitute as it is de-ionised and contains only limited amounts of mineral impurities. If absolutely necessary, tap water can be used, especially if it is known to be of the soft type. Using non-distilled water will inevitably lead to early 'furring-up' of the system and the need for more frequent flushing. The correct water/antifreeze mixture is 50/50; do not allow the antifreeze level to fall below 40% as the anti- corrrosion properties of the coolant will be reduced to an unacceptable level. Anti-freeze of the ethylene glycol based type should always be used. Never use alcohol based anti-freeze in the engine.

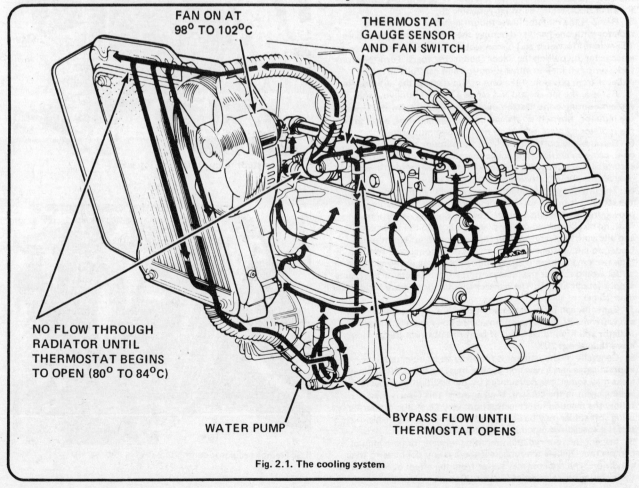

FAN ON AT 98° TO 102°C

THERMOSTAT GAUGE SENSOR AND FAN SWITCH

NO FLOW THROUGH RADIATOR UNTIL THERMOSTAT BEGINS TO OPEN (80° TO 84°C)

WATER PUMP

BYPASS FLOW UNTIL THERMOSTAT OPENS

Fig. 2.1. The cooling system

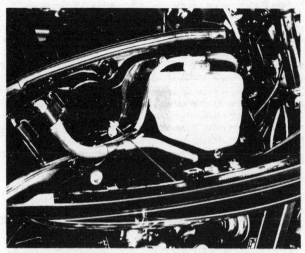

4.1 Radiator reservoir is contained within dummy tank

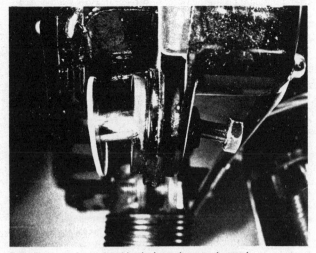

5.3a Radiator is retained by bolts at lower edge and ...

5   Radiator: removal, cleaning and examination

1   Drain the radiator as described in Section 2 of this Chapter.
2   Disconnect the top hose at the thermostat manifold by loosening the screw clip. Because the lower hose is of a short length only, it is easier to disconnect it by detaching the manifold from the water pump casing. The manifold is retained by two bolts. The screw clip can now be loosened and the manifold pulled from the lower hose. Disconnect the reservoir hose from the union on the filler neck. Note the spring retaining clip.
3   Remove the radiator lower mounting bolts. Support the radiator with one hand and remove the top mounting nuts. Ease the radiator forwards and disconnect the electrical leads to the electric fan by pulling the 'block' connector apart. Turn the front forks onto full lock in either direction and carefully lift the radiator from position. Take care not to damage the radiator fins.
4   Remove the fan shroud and fan as a unit by removing the three mounting bolts. Detach the radiator guard from the front of the radiator, where it is retained by four domed nuts, and remove the radiator shrouds, which are held by two nuts each.
5   Remove any obstructions from the exterior of the radiator core, using an air line. The congolmeration of moths, flies, and autumnal detritus usually collected in the radiator matrix severely reduces the cooling efficiency of the radiator.
6   The interior of the radiator can most easily be cleaned while the radiator is in-situ on the motorcycle, using the flushing procedure described in Section 3 of this Chapter. Additional flushing can be carried out by placing a hose in the filler neck and allowing the water to flow through for about ten minutes. Under no circumstances should the hose be connected to the filler neck mechanically as any sudden blockage in the radiator outlet would subject the radiator to the full pressure of the mains supply (about 50 psi). The radiator should not be tested to greater than 15 psi.
7   Bent fins can be straightened, if care is excercised using two screwdrivers. Badly damaged fins cannot be repaired; a new radiator will have to be fitted, if bent fins obstruct the air flow more than about 20%
8   Generally, if the radiator is found to be leaking, repair is impracticable and a new component must be fitted. Very small leaks may sometimes be required by the addition of a special sealing agent in the coolant. If an agent of this type is used follow the manufacturers instructions very carefully. Soldering, using soft solder may be efficacious for caulking large leaks but this is a specialised repair best left to experts.
9   Inspect the four radiator mounting rubbers for perishing or compaction. Renew the rubbers if there is any doubt as to their condition. The radiator may suffer from the effect of vibration if the isolating characteristics of the rubber are reduced.

5.3b ... on studs through frame at top edge

5.3c Remove radiator complete with fan and scoops

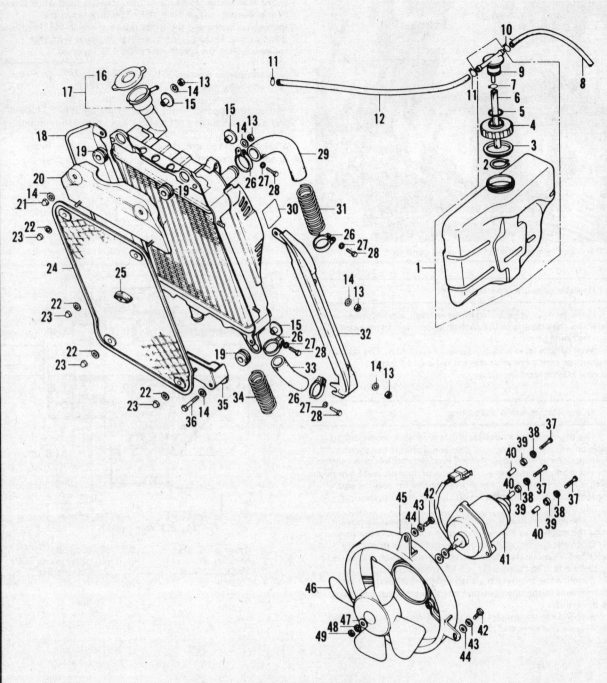

Fig. 2.2. Radiator assembly

1	Reservoir assembly	13	Nut - 8 off	25	Anti-knock grommet	37	Screw - 3 off
2	External circlip	14	Plain washer - 10 off	26	Hose clamp - 4 off	38	Plain washer - 3 off
3	Gasket	15	Spacer - 4 off	27	Spring washer - 4 off	39	Rubber bushing - 3 off
4	Reservoir cap	16	Radiator pressure cap	28	Bolt - 4 off	40	Spacer - 3 off
5	'O' ring	17	Radiator	29	Top hose	41	Fan motor
6	Coolant syphon tube	18	RH air scoop	30	Warning label	42	Bolt - 3 off
7	Hose clip	19	Mounting rubber - 4 off	31	Anti-kink spring	43	Spring washer - 3 off
8	Breather hose	20	Upper guard	32	LH air scoop	44	Plain washer - 3 off
9	Two-way cap	21	Acorn nut - 2 off	33	Lower hose	45	Fan shroud
10	Hose clip	22	Plain washer - 4 off	34	Anti-kink spring	46	Fan blade
11	Hose clip - 2 off	23	Acorn nut - 4 off	35	Lower guard	47	Plain washer
12	Coolant hose	24	Radiator stone guard	36	Bolt - 2 off	48	Spring washer
						49	Nut

5.9 Rubber mountings must be in good condition

to the thermostat. The type of thermometer used when preparing fruit preserves is ideal. Heat the oil , noting when the thermostat opens and the temperature at which the thermostat is fully open. If the performance is at variance with the following table the thermostat should be renewed.

Valve begins opening 80° - 84^oC (176° - 183^oF)
Valve fully open 95^oC (203^oF)

Heat the thermostat for about 5 minutes at 97^oC (207^oF) and measure the valve lift which should be 0.315in (8mm)

4   Refit the thermostat so that the bleed orifice is in the uppermost position.

6   Radiator pressure cap: testing

1   If the valve or valve spring in the radiator cap becomes defective the pressure in the cooling system will be reduced, causing boiling over.
2   Most garages have a special pressure cap tester. The correct pressure at which the valve will lift is 10.7 - 14.9 psi (0.75 - 1.05 hg-·cm^2).

7   Thermostat: removal and testing

1   The thermostat is so designed that it remains in the closed position when it is in a normal cold condition. If the thermostat malfunctions, it will remain closed even when the engine reaches normal working temperature. The flow of coolant will be impeded so that it does not pass through the radiator for cooling and consequently the temperature will rise abnormally, causing boiling over.
2   If the performance of the thermostat is suspect, remove it from the machine as follows and test it for correct operation.
    Drain the coolant and remove the radiator as previously described. Remove the thermostat manifold, which is retained by two bolts. The thermostat will lift from position.
3   Examine the thermostat visually before carrying out tests. If it remains in the open position at room temperature, it should be discarded.
    Suspend the thermostat by a piece of wire in a pan of cold oil. Place a thermometer in the oil so that the bulb is close

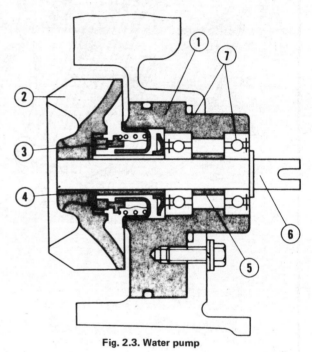

Fig. 2.3. Water pump

1   Water pump body
2   Impeller
3   Mechanical seal
4   Collar A
5   Collar B
6   Water pump shaft
7   Ball bearing

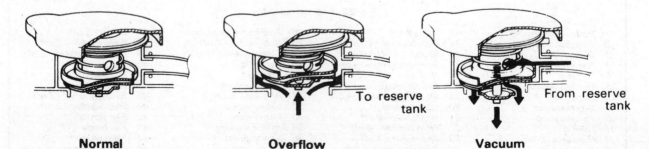

Normal   Overflow   To reserve tank   Vacuum   From reserve tank

Fig. 2.4. Radiator cap: mode of operation

8 Water pump: removal, renovation and replacement

1   Malfunction of the water pump requires renewal of the
component as repair is impracticable. The most likely faults
are looseness and damage to the impeller or leakage of the
oil/water seal leading to failure of the pump bearings.

2   Leakage of the sea will allow small amounts of water to
find its way into the lubrication system or oil seeping into the
coolant. In either event the pump should be renewed
immediately, removal and replacement being carried out as
follows.
    Drain the cooling system and also the engine oil.
Remove the water pump casing, which is retained by 3 or 4
screws. The lower hose manifold need not be detached.
The pump casing is located on two dowels and owing to this
and the type of gasket used, may be difficult to remove.
A rawhide mallet may be used to remove the case. Remove the
radiator as described in Section 5.

3   Examination of the impeller can be made at this stage,
without further dismantling.
    Unscrew the centre bolt and remove the oil filter housing.
Loosen the engine front cover screws evenly and detach the
cover. Note the position of the various 'O' rings, hollow
dowels and collars.

4   The water pump is retained from inside the cover by
three bolts. Remove the bolts and drift the pump from
position. Do not tap the shaft end or damage will result.
Check the bearings for wear and for any evidence of leakage
past the shaft. Seepage of oil or water thought to be caused by
a faulty pump seal may be caused by faulty or omitted
'O' rings though this is not very likely. As mentioned
previously, repair of this component is impracticable and in
any event spares are not available over the counter.

5   Replace the water pump by reversing the order given for
dismantling, checking that all 'O' rings are in good condition and
are not omitted. Coat both sides of the pump casing gasket with
a waterproof gasket compound.

9 Fan motor: testing

Refer to Chapter 7 Section 21

10 Sensor switches: testing

Refer to Chapter 7 Sections 20 and 21.

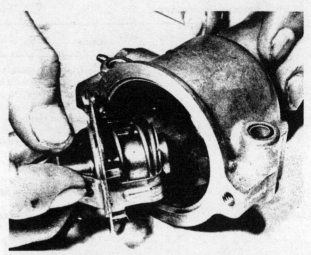

7.4 Thermostat is fitted with bleed orifice upwards

8.2 Water pump casing is located on two dowels

11 Fault diagnosis: Cooling system on page 82

11 Fault diagnosis: cooling system

Symptom	Cause	Remedy
Overheating	Insufficient coolant in system	Top-up reservoir
	Radiator core internally blocked	Flush out system
	Radiator core externally blocked	Remove and clean radiator
	Hoses collapsed blocking flow	Remove and fit new hose(s)
	Thermostat not opening correctly	Remove and fit new thermostat
	Blowing cylinder head gasket (water/steam being expelled into reservoir and out of breather under pressure	Remove cylinder head and fit new gasket
	Low percentage of antifreeze	Drain and refill with 50/50 mixture
	See also mechanical causes fault diagnosis, Chapter one.	
Underheating	Thermostat jammed open	Remove and renew
	Incorrect grade of thermostat	Fit thermostat in correct range
	Thermostat omitted from system	Check and fit correct thermostat
Loss of coolant	Loose hose clips	Tighten clips
	Hose leaking	Renew hose
	Radiator core leaking	Small leaks, use anti-leak compound Large leaks, remove for repair or renewal
	Thermostat, water pump casing or manifold gasket leaking	Renew gasket
	Radiator cap valve spring worn	Test cap and renew
	Blown cylinder head gasket (water/steam being forced out of system underpressure	Remove cylinder head and renew gasket
	Cylinder wall or head cracked	Dismantle engine, renew damaged parts or have expert repair made

Chapter 3 Fuel system and lubrication

Contents

Specifications

Fuel tank capacity	5.0/4.2 US/Imp. gallon (19 litres)
Reserve capacity	0.8/0.7 US/Imp. gallon (3 litres)

Engine oil capacity

After draining	3.2/2.6 US/Imp. quart (3.0 litres)
After dismantling	3.7/3.1 US/Imp. quart (3.5 litres)
Final drive capacity	200 - 220 cc (6.8 - 7.5 fl. oz)

Carburettors

	1975 model	1976 model	1977 model
Make	Keihin	Keihin	Keihin
Type		Constant vacuum (CV) 32 mm	
Primary main jet	65	62	62
Secondary main jet	125	120	125
Pilot fuel jet	35	35	35
Primary air jet	120	120	120
Secondary air jet	60	60	60
Pilot air jet	110	115	110
Air screw no. turns out	1 3/8	2	2½
Float level	0.826 in (21 mm)	0.826 in (21 mm)	0.826 in (21 mm)

Fuel pump

Type	Cam operated, diaphragm

Oil pumps

Main pump type	Earls trochoid
Clutch scavenge pump type	
Rotor/body radial clearance (both pumps)	0.0059 - 0.0083 in (0.15 - 0.21 mm)
Wear limit	0.0161 in (0.41 mm)
Rotor/bodyside clearance	
Main pump	0.0008 - 0.0028 in (0,02 - 0.07 mm)
Wear limit	0.0047 in (0.12 mm)
Clutch pump	0.0008 - 0.0039 in (0.02 - 0.1 mm)
Wear limit	0.0047 in (0.12 mm)
Inner/our rotor radial clearance:	
Both pumps	0.0059 in (0.15 mm)
Wear limit	0.0138 in (0.35 mm)

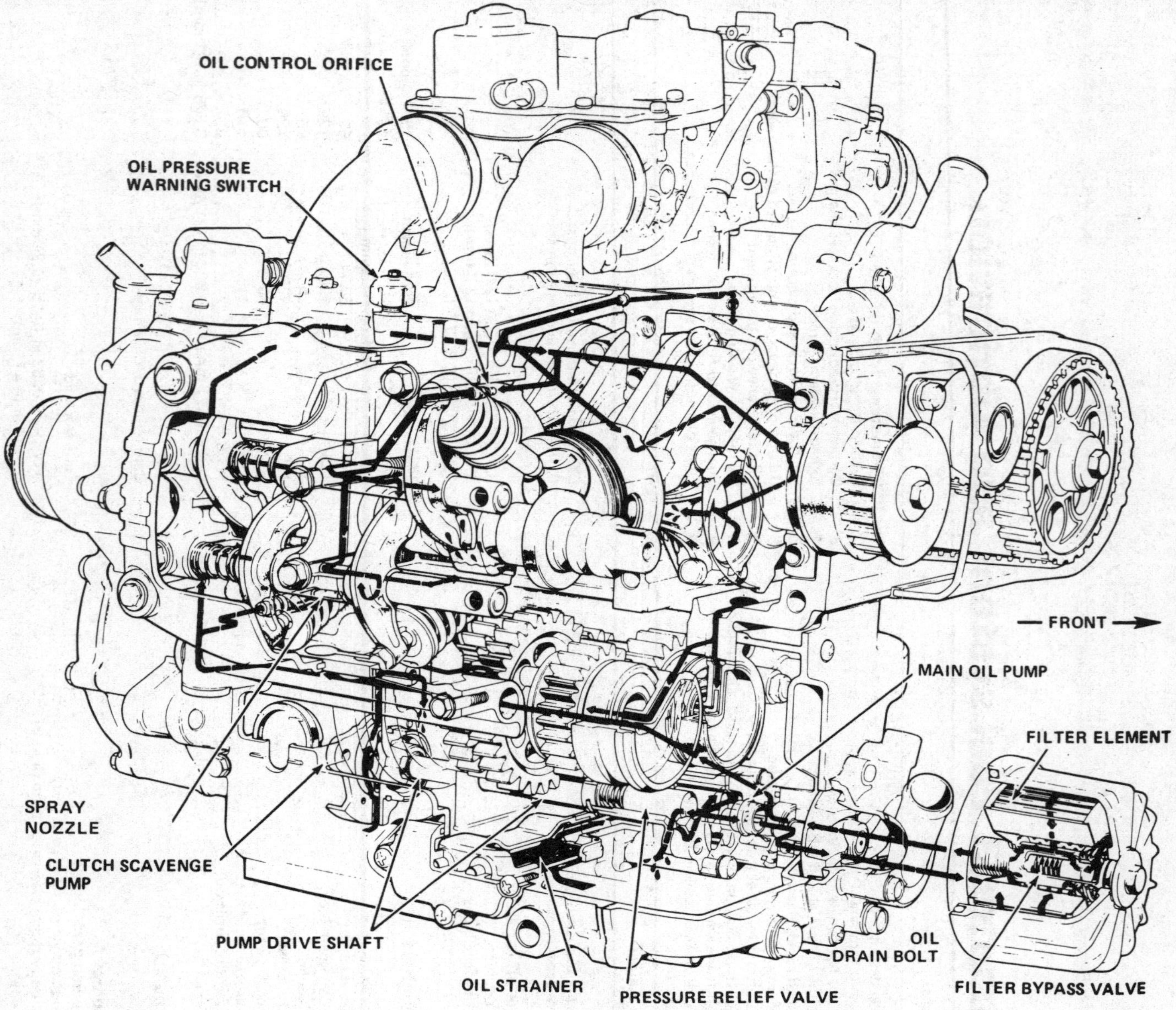

OIL CONTROL ORIFICE

OIL PRESSURE
WARNING SWITCH

SPRAY
NOZZLE

CLUTCH SCAVENGE
PUMP

PUMP DRIVE SHAFT

OIL STRAINER

PRESSURE RELIEF VALVE

OIL
DRAIN BOLT

MAIN OIL PUMP

← FRONT

FILTER ELEMENT

FILTER BYPASS VALVE

Fig. 3.1. The lubrication system

1 General description

1 The fuel system fitted to the Honda Gold Wing is similar to that used on most modern motorcycles. It is, however, unusual in the three following particulars. The fuel is contained within a tank fitted below the dualseat and forward of the rear mudguard the normal position of the fuel tank being occupied by a dummy tank containing the air filter and electrical components. An accurate indication of the quantity of fuel being carried is given by a fuel gauge, mounted in the centre panel of the dummy tank. Fuel is fed from the tank to the carburettors by means of a diaphragm type fuel pump, fitted to the rear of the right-hand cylinder head and operated by a cam lobe extending from the valve operating camshaft. The remainder of the fuel system follows normal motorcycle practice. Four 32 mm Keihin constant vacuum carburettors are mounted over the top of the engine, on a common cast aluminium air box and feeding individual cylinders through separate inlet manifolds. The air box is attached to an air filter box within the dummy fuel tank, containing a corrugated paper type element. The carburettors are interconnected by a control rod and are operated from a throttle twist grip by push-pull cables around a single pulley. Although a fuel pump is fitted, a fuel tap is incorporated in the system, attached to the fuel tank, from where it supplies the fuel through a hose and filter to the intake side of the pump. The tap incorporates a 'Reserve' position to enable an additional supply of 0.8/0.7 US/Imp. quart (3.0 litres) of fuel to be available when the main supply runs out. For cold starting, a cable-operated choke is fitted, attached to rear right-hand carburettor and interconnected with the remaining three instruments. The choke operating knob is located to the left of the lighting console, below the speedometer. Lubrication is by the wet sump principle in which oil is delivered, under pressure, from the sump by a mechanical pump to the working parts of the engine. Oil is returned to the sump by gravity and by a secondary pump which scavenges oil trapped in the clutch housing. The two pumps are driven by a shared shaft running the length of the engine and driven by a duplex chain taken off a sprocket to the rear of the clutch outer drum. Both pumps are of the trochoid rotating vane type. Oil is picked up by the main pump, through a gauze strainer in the sump and passed under pressure through a full flow oil filter fitted with a paper element. The engine oil supply is also shared by the gearbox and primary drive.

2 Fuel tank: removal

1 The fuel tank is retained in position by a single bolt through a bracket at the upper rear edge of the tank, and is supported on two rubber cushions by brackets projecting from the frame.
2 Removal is unnecessary except when the tank itself has become damaged or attention to the rear portion of the frame is required. Owing to the unusual shape of the tank and its close proximity to the frame members and ancillary components bolted to the frame removal is a laborious process.
3 Commence tank removal by detaching the dualseat and the side panels. Follow this by removing the rear wheel, mudguard and any other component which may obstruct the tank as it is lifted out towards the rear. Disconnect the fuel line at the fuel pump and allow the tank to drain. Disconnect the two leads from the fuel gauge sensor unit.
4 Close inspection of the machine will reveal why removal of the tank is so difficult, and should not be undertaken unless absolutely necessary.

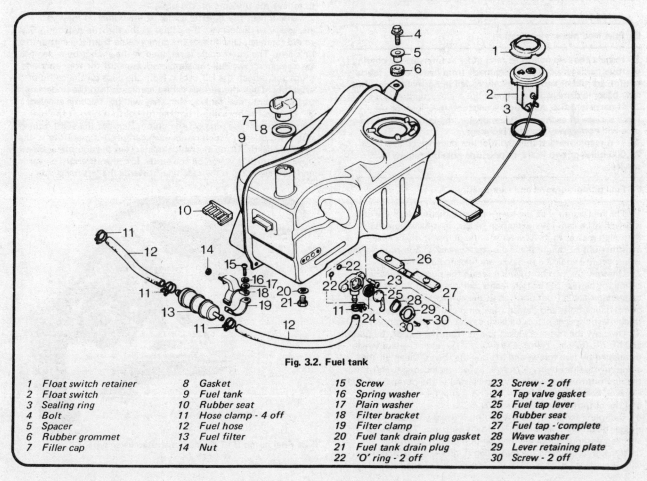

Fig. 3.2. Fuel tank

1	Float switch retainer	8	Gasket	15	Screw	23	Screw - 2 off
2	Float switch	9	Fuel tank	16	Spring washer	24	Tap valve gasket
3	Sealing ring	10	Rubber seat	17	Plain washer	25	Fuel tap lever
4	Bolt	11	Hose clamp - 4 off	18	Filter bracket	26	Rubber seat
5	Spacer	12	Fuel hose	19	Filter clamp	27	Fuel tap - 'complete
6	Rubber grommet	13	Fuel filter	20	Fuel tank drain plug gasket	28	Wave washer
7	Filler cap	14	Nut	21	Fuel tank drain plug	29	Lever retaining plate
				22	'O' ring - 2 off	30	Screw - 2 off

3 Fuel tap: removal and replacement

1 Fuel tap removal is seldom necessary unless the joint between the tap and tank starts to leak or the valve rubber gasket fails. Before removing the tap or lever, the tank must be drained of fuel. Detach the feed hose at the pump and feed the fuel into a suitable clean container.

2 Failure of the valve gasket can be remedied without removing the tap body from position. Unscrew the lever retaining plate and detach the wave spring and lever, followed by the rubber gasket. The new gasket can be fitted and the outer components replaced.

3 To remove the complete tap, loosen the screw clip from around the feed hose and pull the hose from the tap union. The tap body is held in position by two screws passing through a flange into the tank. Remove the screws and detach the body. Note the two 'O' rings which seal the flange/tank joint. These should be replaced by new ones when the tap is disturbed.

4 Fuel filter: removal and replacement

1 In addition to two fuel strainers which are fitted in the bottom of the fuel tank, a fuel filter is fitted midway in the tank to fuel pump line.

2 It is recommended that the filter is renewed every 24,000 miles or two years. The filter is a sealed unit and therefore cleaning is impracticable. A new unit must be fitted.

3 The filter is secured by a clamp bracket, which is retained by one bolt to the lower seam of the fuel tank. Remove the filter by separating the clamp and pulling the fuel feed hose off each end of the filter, after loosening the screw clips.

5 Fuel feed pipes: examination

1 Inspect the two main fuel feed pipes at intervals for chafing or other damage. Both pipes are made from heavy duty, fabric reinforced rubber and are unlikely to fail in normal service, but may become damaged if the hose clips are overtightened.

2 If persistent fuel starvation is encountered, inspect the inner surface of each pipe. Occasionally, the inner liner of the pipe will come away, causing blockage.

3 It is recommended that the fuel line be renewed every 24,000 miles or two years, to preclude possible failure in service.

6 Fuel pump: removal and examination

1 The fuel pump is of the lever-operated diaphragm type and is driven off a cam lobe extension of the right-hand camshaft. If foreign matter finds its way into the pump, it may cause malfunctions of the intake and output valves. If pump performance is suspect, it should be removed and cleaned as follows.

2 Unscrew the two bolts which retain the pump mounting casing to the rear of the right-hand cylinder head. Pull the tachometer cable from position in the casing. Remove the two pump flange bolts and detach the pump complete. Note the thick plastic spacer pieces on each casing mounting flange.

3 Unscrew the five cross-head screws from the top of the pump and lift off the cap. Remove the valve cage retainer plate which is retained by two screws and lift out the valves. Clean all the components thoroughly in petrol, before replacement. Refit the non-return valves so that when viewed in the pumps' normal working position the valve on the input side faces downwards and the output valve faces upwards.

4 Complete failure of the fuel pump is usually caused by the rubber diaphragm perishing and finally splitting. This can be checked without removing the fuel pump body from the machine, by detaching the pump cap. Failure of the diaphragm will be evident on inspection. Unfortunately, owing to modern

design trends, the diaphragm cannot be detached from the pump as a whole. Nor are spare diaphragms available as spares. The complete unit must be renewed.

5 Reassemble the pump in the reverse order given for dismantling. If necessary, renew the gaskets which lie either side of the mounting casing insulator piece.

6 If required, the performance of the fuel pump can be checked as follows. Detach the hose from the output side and reconnect the hose via a T-piece. Attach a fluid pressure gauge to the T-piece by means of a length of hose. Run the engine at the following speeds and compare the pressure readings recorded with those below

500 rpm	2.4 psi
900 rpm	2.3 psi
5000 rpm	2.0 psi

The flow rate of the pump can also be checked by using the same T-piece, the hose from which leads into a measuring jug. Run the engine at a speed of 3000 rpm. The pump should supply 0.475 Imp. Quart (0.45 litres/15 Fl.oz) per minute. The tests described above are usually unnecessary as examination of the pump will reveal any irregularities.

7 Carburettors: removal from the machine

1 Open the three panels of the dummy fuel tank and remove the tool tray. Remove the air filter box cover and lift out the filter element. The air filter box can be lifted from position after unscrewing the two bolts which hold it to the carburettor air box, and detaching the breather pipe. The pipe is retained by a spring clip, the 'ears' of which should be pressed together to relieve the pressure on the pipe.

2 Disconnect the throttle cables at the pulley. It may be necessary to disconnect the cables at the throttle twist grip first, to aid removal. Disconnect the choke cable from the carburettor link arm. The outer cable is retained at the anchor by a clamp held by a single screw. Pull the suppressor caps off the four sparking plugs and detach the HT cables from the clips on the carburettor assembly. Tuck the various cables and leads into the underside of the dummy fuel tank so that they will not become tangled.

3 Loosen the hose clip from the fuel output pipe at the fuel pump and pull the pipe off the union. Remove the eight domed nuts, two of each which retain each carburettor manifold to the cylinder heads. Using a rawhide mallet, tap the complete assembly upwards, until it is free on the studs. Lift the assembly upwards and out, away from the machine, towards the left-hand side.

6.2a Fuel pump is of the cam operated lever type

6.2b Retained by flange bolts to the tachometer drive casing

6.3 Valves must be fitted as shown

7.3 Carburettors removed from machine as a complete unit

8 Carburettors: dismantling examination and reassembly

1   Removal of the carburettors from the air box need not take place unless the instruments are to be renewed or the air box itself requires renewal. Follow the procedure below for carburettor removal. Prise the 'E' clips off the link pivots on the end of the throttle control rods on No. 3 and No. 4 carburettors. Bend the detent bolt tab washers down and remove the detent bolt, spring and plunger from the outer end of each link. Detach the link ends of the control rods from the pivots. Remove the five cross-head screws which hold the two halves of the air box together. These screws may be very tight, having been assembled using a locking fluid. Great care should be taken as the screws are of a soft material and will shear easily. After removal of the screws prise the split pins or spring pins from the choke control rods and detach them from the links, as the two halves of the air box are separated. Do not prise the air box halves apart using screwdrivers or other levers as the sealing ring or mating surfaces may be damaged.

2   The two pairs of carburettors can now be removed from their respective air box halves, using an identical procedure for each as follows' Press the 'ears' of the slow air hose clips together and pull the hoses off the unions with the air box. Bend down the carburettor mounting bolt tab washer ears and remove the bolts. Pull the two carburettors off the air box as a pair. Separate the carburettors by pulling them apart, taking care not to lose the springs placed between the throttle and choke link connectors.

3   The procedure set out for removal and separation of the carburettors in the previous paragraphs need not take place for normal examination and removal of the carburettor internal components such as the float assembly and jets. It is strongly advised that each carburettor be dismantled and reassembled separately, to prevent accidently interchanging the components. Dismantle and examine each carburettor, following an identical procedure as described below.

4   Detach the slow air hose from one carburettor at the union on the carburettor body. Unscrew the slow air jet. When removing any jet from a carburettor ensure that the screwdriver is of the correct size, fitting the slot closely. This will prevent damage to the jet and prevent burring, which may alter the orifice size.

5   Invert the carburettor assembly and remove the four screws which retain the float bowl to the carburettor. Lift the float bowl from position, and remove the gasket. Gently pull on the metal tab to remove the primary and secondary main jets. The jets are push fit located and sealed by small 'O' rings. The two floats, which are interconnected, can be lifted from position after pushing the pivot pin out of the pivot posts. The float needle will come away with the float assembly as it is retained by a spring clip hooked around the float tongue. Pull out the needle valve seat after removing the retaining claw (tongue) which is retained by a single screw.

6   Carefully prise the rubber bung from position between the main jet housings and unscrew the pilot jet, which is located below the bung. The needle jet and main nozzle, which lie in the main jet housings, can be removed by pushing them out from the venturi side of the carburettor, using a finger or a suitable piece of wood which will not damage the brass. Removal of these two components is probably easier to accomplish after detaching the carburettor cap and piston as follows.

7   Lift the carburettor cap off the main body, after removing the three retaining screws. Pull out the piston very carefully, so that the piston needle does not get bent. The needle can be removed after lifting out the helical spring and removing the grub screw from the piston centre tube. If the piston is inverted the needle will fall out. Two air jets are obscured by a small curved plate and gasket, which is retained by a single screw. Remove the plate, followed by the air jets.

8   It is not recommended that the 'butterfly' valves of either the throttle or choke be removed. The valves themselves are not subject to wear. If wear occurs on the operating pivots, a new

carburettor will be required, as air will find its way along the pivot bearings, resulting in a weak mixture.

9   Check the condition of the floats. If they are damaged in any way, they should be renewed. The float needle and needle sealing will wear after lengthy service and should be inspected closely. Wear usually takes the form of a groove or ridge, which will cause the float needle to seat imperfectly. Always renew the seating and needle, as a pair. An imperfection in one component will soon produce similar wear in the other. The needle seat is fitted with a fine filter gauze. Renew if holed.

10   After considerable service, the piston needle and the needle jet in which it slides, will wear, resulting in an increase in petrol consumption. Wear is caused by the passage of petrol and the two components rubbing together. It is advisable to renew the jet periodically in conjunction with the needle. The vacuum piston and carburettor cap also work as a pair. Examine the components for scoring and other damage, checking particularly that the piston does not have a 'tight spot' anywhere in its travel. Such a 'tight spot' could be caused by a bent needle so carry out the check with the needle detached. Never interchange the piston or cap of one carburettor with that of another.

11   Before the carburettors are reassembled, each should be cleaned out thoroughly, using compressed air. Avoid using a piece of rag since there is always risk of particles of lint obstructing the airways or jet orifices. Never use a piece of wire or any pointed metal object to clear a blocked jet. It is only too easy to enlarge a jet under these circumstances and increase the rate of petrol consumption. If an air line is not available, a blast of air from a tyre pump will usually suffice.

12   Reassemble each carburettor, using the reversed dismantling procedure. Work must be carried out in absolute cleanliness. If possible, use new gaskets, 'O' rings can be reused if there is no doubt as to their condition. When replacing the piston, note that it can be refitted in only one position. The groove in the piston side must locate with the projection in the main body. Assemble the carburettors in their pairs on the air box halves and then screw the halves together. Reconnect the choke operating rod and the throttle control rods. Remember to secure the detent spring bolts with the tab washers. Check that when the choke mechanism is operated, the choke valves are fully closed. If necessary, adjust the linkage by means of the screw on the nipple anchor of No. 3 carburettor.

13   Do not use excessive force when reassembling a carburettor since it is easy to shear a jet or some of the smaller screws. Furthermore, the carburettors are die cast in a zinc-based alloy which itself does not have a high tensile strength.

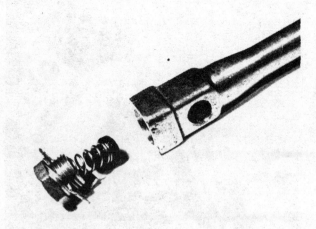

8.1b ... remove detent assemblies to allow link rod removal

8.1c Take care not to damage sealing ring when separating

8.1a Prise off the 'E' clip from the pivots and ...

8.2a Bend down tab washers to remove carburettors

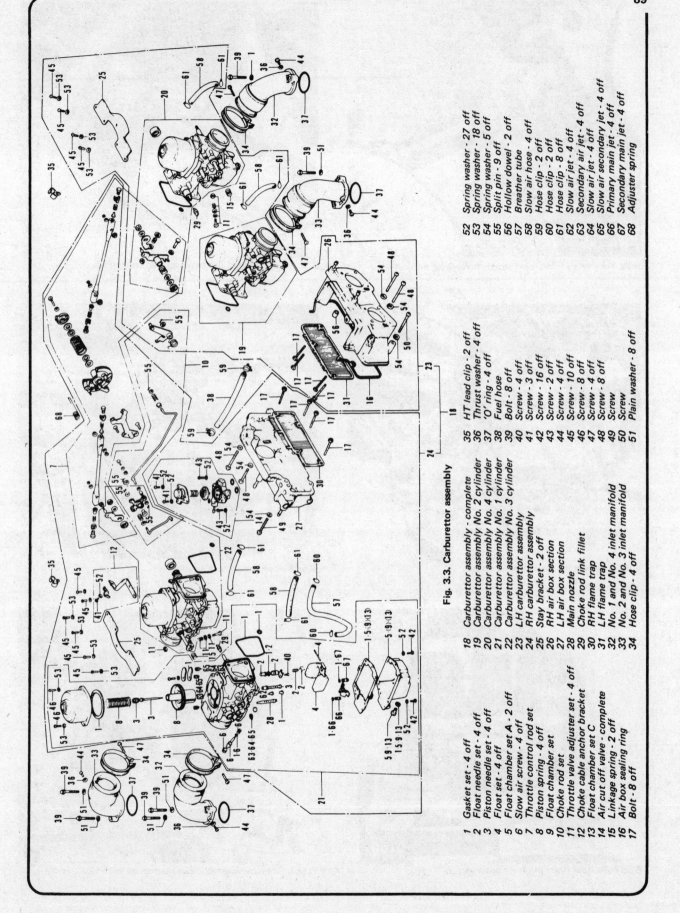

Fig. 3.3. Carburettor assembly

1 Gasket set - 4 off
2 Float needle set - 4 off
3 Piston needle set - 4 off
4 Float set - 4 off
5 Float chamber set A - 2 off
6 Slow air screw - 4 off
7 Throttle control rod set
8 Piston spring - 4 off
9 Float chamber set
10 Choke rod set
11 Throttle valve adjuster set - 4 off
12 Choke cable anchor bracket
13 Float chamber set C
14 Air cut off valve - complete
15 Linkage spring - 2 off
16 Air box sealing ring
17 Bolt - 8 off

18 Carburettor assembly - complete
19 Carburettor assembly No. 2 cylinder
20 Carburettor assembly No. 4 cylinder
21 Carburettor assembly No. 1 cylinder
22 Carburettor assembly No. 3 cylinder
23 LH carburettor assembly
24 RH carburettor assembly
25 Stay bracket - 2 off
26 RH air box section
27 LH air box section
28 Main nozzle
29 Choke rod link fillet
30 RH flame trap
31 LH flame trap
32 No. 1 and No. 4 inlet manifold
33 No. 2 and No. 3 inlet manifold
34 Hose clip - 4 off

35 HT lead clip - 2 off
36 Thrust washer - 4 off
37 'O' ring - 4 off
38 Fuel hose
39 Bolt - 8 off
40 Screw - 4 off
41 Screw - 3 off
42 Screw - 16 off
43 Screw - 2 off
44 Screw - 4 off
45 Screw - 10 off
46 Screw - 4 off
47 Screw - 4 off
48 Screw - 8 off
49 Screw
50 Screw
51 Plain washer - 8 off

52 Spring washer - 27 off
53 Spring washer - 18 off
54 Spring washer - 5 off
55 Split pin - 9 off
56 Hollow dowel - 2 off
57 Breather tube
58 Slow air hose - 4 off
59 Hose clip - 2 off
60 Hose clip - 2 off
61 Hose clip - 8 off
62 Secondary air jet - 4 off
63 Secondary air jet - 4 off
64 Slow air jet - 4 off
65 Slow air secondary jet - 4 off
66 Primary main jet - 4 off
67 Secondary main jet - 4 off
68 Adjuster spring

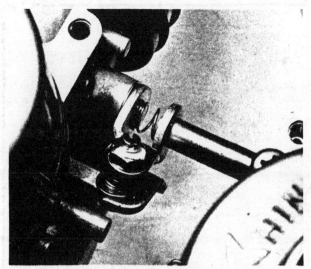
8.2b Take care not to lose throttle link spring and ...

8.2c ... connecting fillet between two choke rods

8.4 Disconnect slow air hoses at unions

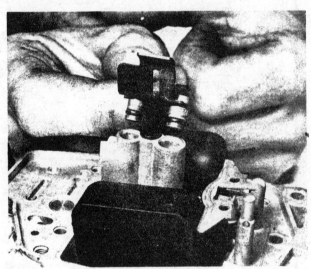
8.5a Pull on tab to remove main jets

8.5b Push float pivot pin from posts and ...

8.5c ... lift float assembly away complete with float needle

8.5d Remove screw and claw to ...

8.5e ... free float valve seat from housing

8.6a Prise out rubber bung to ...

8.6b ... gain access to the pilot jet

8.6c Push jet nozzle out from venturi side

8.6d Blanking plug will fall out

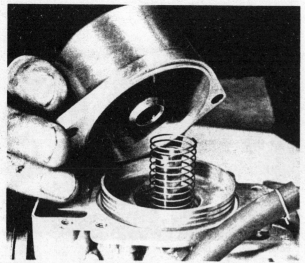

8.7a Lift carburettor cap off and ...

8.7 b ... pull piston and needle out of bore

8.7c Piston needle retained by a grub screw

8.7d The two air jets are obscured by ...

8.7e ... a plate and gasket retained by a screw

8.12 Main jet plate must be replaced as shown

8.13a Carburettor linkage, general view

8.13b Carburettor linkage, general view

8.13c Ensure all seals and 'O' rings are perfect

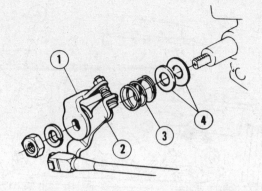

Fig. 3.4. Connecting throttle lever linkages

1	Link lever	3	Spring
2	Throttle lever	4	Washer

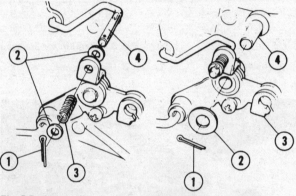

Fig. 3.5. Connecting the choke rod assembly

1 Split pin	3 Spring
2 Washer	4 Choke rod

Fig. 3.6. Entering the choke linkage

1 Split pin	3 Choke link
2 Washer	4 Body

9 Carburettor assembly: Air cut-off valve examination

1 An air cut-off valve is fitted to the carburettor box and interconnected to all the carburettors. The valve is of the diaphragm type and automatically regulates the amount of air travelling through the pilot air system of the carburettors. When the engine is running at idling speeds, the valve remains open, allowing the correct quantity of air to enter the pilot system. If the throttle is closed when the engine has been running fast, the high vacuum in the inlet manifolds causes the valve to close, reducing the air flow and creating a rich mixture to prevent backfiring.

2 The valve assembly may be dismantled without removing the carburettors from the machine. The air filter box should be removed to give improved access. Disconnect the air cut-off valve main hose and remove the two screws which retain the complete assembly to the mounting point on the air box.

3 Remove the valve cover, which is retained by two screws and lift the cover from position. Lift out the helical spring and valve diaphragm. Clean all the components in petrol. The valve plate and plate retainer will remain in the main valve body and should not be removed.

4 Check the diaphragm for splitting or other damage. Use compressed air to clean the by-pass channel and the valve plate seat. Inspect the two larger 'O' rings and the small 'O' ring that fits in the by-pass orifice. Renew components as necessary.

5 Reassemble the components carefully and refit the completed unit to the air box.

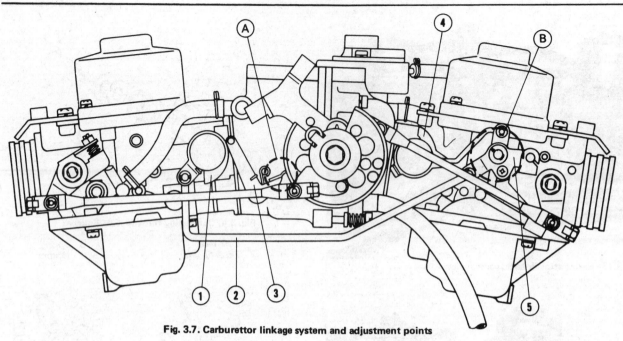

Fig. 3.7. Carburettor linkage system and adjustment points

1	L-choke lever	3	Fast idle link	5	Choke link	B	Choke synchronizing
2	Choke rod	4	R-choke lever	A	Fast idle adjustment point		adjustment point

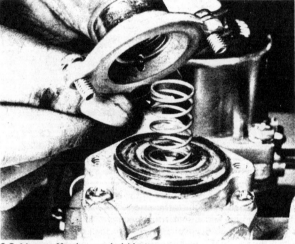

9.2 Air cutoff valve cap held by two screws

10 Carburettor: Idle speed adjustment

1 Adjustment of tick-over on most motorcycles is made by careful adjustment of the pilot air screws in conjunction with the throttle stop screw(s). On the Honda Gold Wing the pilot air screw on each carburettor is adjusted at the factory and should not be altered unless new pilot screws are being fitted. Tick-over adjustment is therefore limited to alteration of the throttle stop screw located immediately below the throttle operating pulley.
2 The adjustment of engine speed should be made only when the engine has reached normal working temperature. The correct tick-over speed is 900 rpm.

11 Carburettors: synchronisation

1 For the best possible performance it is imperative that the carburettors are working in perfect harmony with each other. If the carburettors are not synchronised, not only will one cylinder be doing less work, at any given throttle opening, but it will also in effect have to be carried by the other cylinders. This will reduce performance accordingly.
2 For synchronisation, it is essential to use a vacuum gauge set consisting of four separate dial gauges, one of which is connected to each carburettor by means of a special adaptor tube. The adaptor pipe screws into the outside lower end of each inlet manifold, the orifice of which is normally blocked off by a cross-head screw plug. Most owners are unlikely to posses the necessary vacuum gauge set, which is somewhat expensive and is normally held by Honda service agents who will carry out the synchronisation operation for a nominal sum.
3 If the vacuum set is available to the owner, the adjustment necessary for synchronisation should be made as follows, in the four recommended stages. Place the vacuum gauge set on the machine so that the dials can be easily read. The usual position is between the handlebars. Remove the four blanking plugs from the manifolds and fit the adaptor pieces. Connect the gauges as follows for ease of observation.

No. 1 cylinder	right-hand gauge
No. 2 cylinder	inner right-hand gauge
No. 3 cylinder	inner left-hand gauge
No. 4 cylinder	left-hand gauge

Start the engine allowing it to warm up, and set the speed to about 1000 rpm by means of the throttle stop screw.
4 First adjust No. 1 and No.3 carburettors by means of the adjustment screw which is between the instruments. Loosen the locknut first. Moving the screw in one direction will raise the vacuum on one carburettor and reduce it on the other. Alter the setting until the dial readings are 21 ± 2.5 cm Hg.
5 Adjust No. 2 and No. 4 carburettors in a similar manner, until the vacuum pressure reading is as above. The adjustmen screw on the outside of No. 4 carburettor can now be used to regulate the pressures between the two sets of carburettors, in the same way as the previous adjustments regulated the pressure between single carburettors in the same pair.
6 Set the tick-over to 900 rpm and stop the engine. Remove the adaptors, pipes and gauges and refit the manifold blanking plugs.

12 Carburettors: settings

1   Some of the carburettor settings, such as the sizes of the needle jets, main jets and needles etc, are determined by the manufacturer. Under normal circumstances it is unlikely that these settings will require modification, even though there is provision made. If a change appears necessary it can often be attributed to a developing engine fault.

2   It is recommended the changes in carburettor specifications are not made as the manufacturer will have spent considerable research time, i e. assessing performance of the engine and will have come to a satisfactory conclusion. The only condition where settings may required alteration are when the machine is to be used at very high altitude when touring the more mountainous parts of the globe. In these cases it is probable that the main jet sizes will require reducing to correspond with the change in atmosphere. Consult a Honda specialist before any such alterations are made.

13 Carburettors: adjusting float level height

1   If persistent flooding or fuel starvation is encountered and cannot be attributed to normal causes, the float height on the offending carburettor should be checked and adjusted, if necessary.

2   The carburettor assembly must be removed from the machine and the float bowl of the carburettor to be adjusted must be detached. With the float assembly positioned so that the float needle valve is closed, measure the distance between the raised edge of the float bowl mating surface on the main body and the lower surface of the float. The correct distance is 0.826in (21mm). Make any adjustment by bending the float connecting arm, at the point where the two holes are drilled. Use a pair of electricians pliers to carry out the somewhat delicate adjustment.

14 Exhaust system: removal

1   Unlike the two-stroke engine, the exhaust gases of a four-stroke engine are usually not of an oily nature. The Honda Gold Wing silencer is therefore not fitted with detachable buffles. If any fault develops in the silencer the complete unit should be renewed.

2   Removal of the exhaust system may be carried out as follows. Loosen the silencer clamps at the silencer/exhaust pipe joints. The socket screws which hold the clamps may be caulked, thereby preventing insertion of a suitable socket key. The caulking can be removed using a suitable pointed instrument. Loosen the screws off as much as possible as the clamps are rounded over at the front edge to help secure the pipe ends. Loosen and remove the four nuts which retain each exhaust pipe assembly to the cylinder heads, and slacken off the silencer bracket/pillion footrest mounting bolts. Tilt the silencer downwards at the front and pull the exhaust pipe from the cylinder heads. The exhaust pipes can now be detached individually from the silencer.

3   Support the silencer and remove the retaining bolts. Lower the silencer onto the floor. If the silencer is to be discarded, probably the easiest method of disentangling it from the machine is to push the machine off the centre stand and wheel it forwards over the curved portion of the silencer. Alternatively, the silencer can be removed by carefully easing it around the rear wheel.

4   When replacing the exhaust system, always use new exhaust rings in the ports to prevent gas leakage.

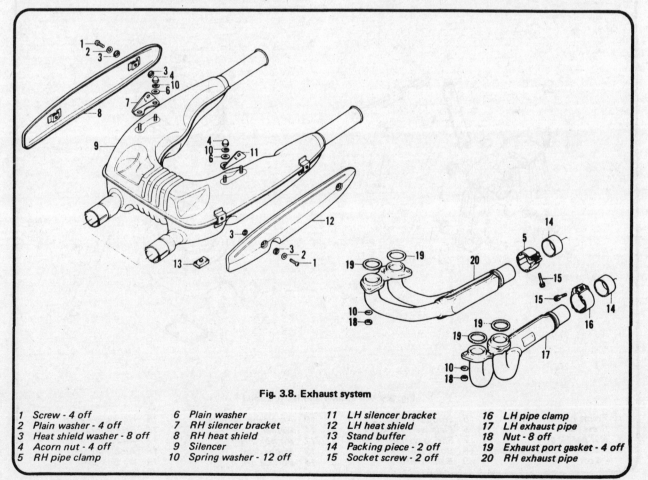

Fig. 3.8. Exhaust system

1 Screw - 4 off	6 Plain washer	11 LH silencer bracket	16 LH pipe clamp
2 Plain washer - 4 off	7 RH silencer bracket	12 LH heat shield	17 LH exhaust pipe
3 Heat shield washer - 8 off	8 RH heat shield	13 Stand buffer	18 Nut - 8 off
4 Acorn nut - 4 off	9 Silencer	14 Packing piece - 2 off	19 Exhaust port gasket - 4 off
5 RH pipe clamp	10 Spring washer - 12 off	15 Socket screw - 2 off	20 RH exhaust pipe

15 Air filter: removal and cleaning

1   Open all three of the dummy tank sections and lift the tool tray from position. Slacken the air filter cover wing nut and remove it and the cover. Lift the air filter element from position.

2   Tap the element gently to remove any loose dust and then use an air hose to remove the remainder of the dust. Apply the air current from the inside of the element only. If an air hose is not available, a tyre pump can be utilised instead. If the corrugated paper element is damp or oily or beginning to disintegrate, it must be renewed.

3   Do not run the engine with the element removed as the weak mixture caused may result in engine overheating and damage to the cylinders and pistons. A weak mixture can also result if the rubber sealing rings on the element are perished or omitted.

4   When replacing the filter assembly, note that the pressed steel cover can only be fitted with the arrow pointing forwards.

16 Oil pumps: removal examination and replacement

1   Unless lubrication failure occurs, the oil pumps should be checked whenever an engine overhaul is undertaken. The main oil pump can be removed from the front of the engine whilst the engine is still in the frame, after following the directions for radiator removal and water pump removal in Sections 5 and

8.2 respectively of Chapter 2. The three pump retaining screws can then be removed and the pump pulled from position, together with the drive shaft. Removal of the clutch scavenge pump can only take place after the engine has been removed from the frame and the clutch itself has been removed. Refer to Chapter 1 for engine removal and dismantling.

2   Dismantle the two pumps separately, cleaning them thoroughly in petrol and allowing them to dry before inspection is carried out. Check the castings for signs of cracking or fracture and inspect the inner wall of the main bodies for scoring. Using a feeler gauge, check the side clearances and radial clearance of each pump and rotors, which sould fall within the specifications as outlined at the beginning of the Chapter. Rotors must be renewed as a set and if the pump body is worn, the complete assembly must be renewed.

3   Examine the rotors and pump bodies for signs of scoring chipping or other surface damage, which will occur if metallic particles find their way into the oil pump assembly. Renewal of the affected parts is the only remedy under these circumstances.

4   Reassemble the pumps in absolutely clean conditions. Even a small particle of grit or metal may damage the rotors. Remember that the punch-marked faces of the rotors must face away from the main pump body. Replace the rotors and lubricate them thoroughly before refitting the cover.

5   Replace the pumps in the engine by reversing the dismantling procedure. Ensure that the 'O' rings fitted to the pumps are in good condition.

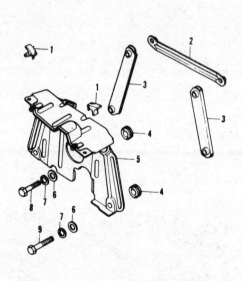
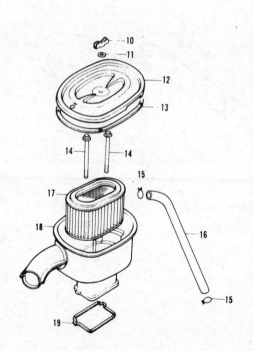

Fig. 3.9. Air cleaner and front shroud

1	Frame rubber - 2 off	5	Shroud
2	Cross bar	6	Plain washer - 4 off
3	Nut bar - 2 off	7	Spring washer - 4 off
4	Rubber mounting grommet	8	Bolt - 2 off
	- 4 off	9	Bolt - 2 off

10	Air cleaner wing nut	15	Hose clip - 2 off
11	Washer	16	Breather hose
12	Air box cover	17	Air cleaner element
13	Rubber seal	18	Air cleaner box
14	Bolt - 2 off	19	Air box seal

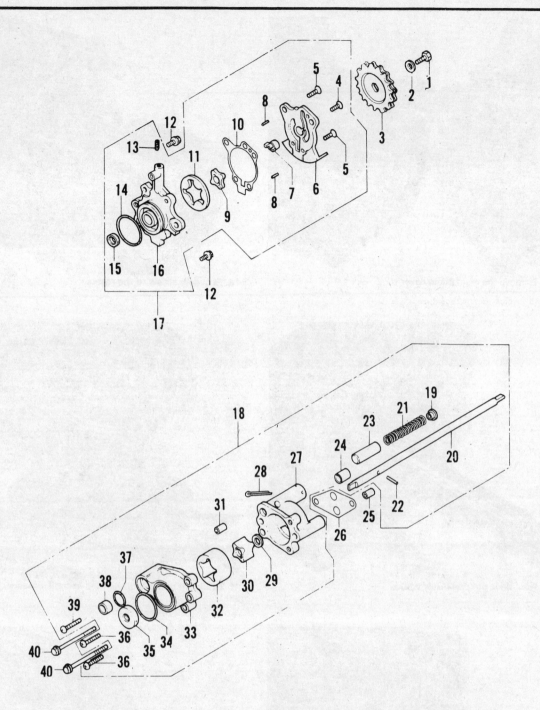

Fig. 3.10. Main oil pump, clutch scavenge pump

1	Bolt	11	Outer rotor	19	Spring seat	30	Inner rotor
2	Washer	12	Bolt - 3 off	20	Pump driveshaft	31	Hollow dowel
3	Pump driven sprocket	13	Grub screw	21	Relief valve spring	32	Outer rotor
4	Countersunk screw	14	'O' ring	22	Drive pin	33	Pump outer body
5	Countersunk screw - 2 off	15	Oil seal	23	Relief valve	34	'O' ring
6	Clutch scavenge pump cover	16	Clutch scavenge pump case	24	Hollow dowel	35	Oil seal
7	Drive collar	17	Clutch scavenge pump - complete	25	Hollow dowel	36	Screw - 2 off
8	Dowel pin - 2 off			26	Gasket	37	'O' ring
9	Inner rotor	18	Main oil pump - complete	27	Oil pump case	38	Feed collar
10	Gasket			28	Split pin	39	Screw - 3 off
				29	Thrust washer	40	Bolt - 2 off

16.2a Remove the main pump cover to ...

16.2b ... gain access to the rotors

16.2c Note drive pin and thrust washer

16.2d Clutch pump is dismantled in similar manner

16.3a Measure rotor/body clearance and ...

16.3b ... inner rotor/outer rotor clearance

16.3c Check all rotors for scoring or chipping

16.3f ... the main pump pressure release valve

16.3d Remove split pin to allow ...

16.3e ... removal of spring and seat of ...

17 Oil filter: renewing the element

1   The oil filter is contained within a separate housing, bolted to the front engine cover by a single hollow bolt containing the pressure release valve. Access to the element is made by unscrewing the centre bolt, which will bring with it the housing and element. Before removing the housing, place a receptacle beneath the engine to catch the engine oil contained in the filter chamber.
2   When renewing the filter element, it is wise to renew the housing 'O' ring at the same time. An 'O' ring is usually provided with a filter.
3   The filter by-pass valve (pressure release valve), comprising a plunger and spring, is situated in the bore of the centre bolt. It is recommended that the by-pass valve be checked for free operation at regular intervals. The spring and plunger are retained by a pin across the centre bolt. Tap the pin out, to allow removal of the plunger and spring for cleaning.
4   Never run the engine without the filter element or increase the period between the recommended oil changes or oil filter changes. Change the engine oil and filter every 6,000 miles or 6 months, whichever is soonest.

18 Oil pressure warning lamp

1   An oil pressure warning lamp is incorporated in the lubrication system to give immediate warning of excessively low oil pressure.
2   The oil pressure switch is screwed into the top of the crankcase, below No. 3. Carburettor. The switch is connected to a warning light on the lighting panel on the handlebars. The light should be on when the ignition is switched on but will usually go out almost as soon as the engine is started.
3   If the oil warning lamp comes on whilst the machine is being ridden, the engine should be switched off immediately, otherwise there is a risk of severe engine damage due to lubrication failure. The fault must be located and rectified before the engine is re-started and run, even for a brief time. Machines fitted with plain shell bearings rely on high oil pressure to maintain a thin oil film between the bearing surfaces. Failure of oil pressure will cause the working surfaces to come into direct contact, causing overheating and eventual engine seizure.

19 Fuel gauge and float switch: testing

Refer to Chapter 7. Section 22.

20 Fault diagnosis: fuel system and lubrication

Symptom	Cause	Remedy
Engine gradually fades and stops	Fuel starvation	Check vent hole in filler cap, or breather tube. Sediment in filter bowl or float chamber. Dismantle and clean
Engine runs badly. Black smoke from exhausts	Carburettor flooding	Dismantle and clean carburettor. Check for punctured float or sticking float needle Check fuel level in float chamber
	Air filter blocked	Remove and clean
Engine lacks response and overheats	Weak mixture Air cleaner disconnected or hose split Modified silencer has upset carburation	Check for partial block in carburettors. Reconnect or renew hose Replace with original design
Oil pressure warning light comes on	Lubrication system failure	Stop engine immediately. Trace and rectify fault before re-starting
Engine gets noisy	Failure to change engine oil when recommended	Drain off old oil and refill with new oil of correct grade. Renew oil filter element

Chapter 4 Ignition system

Contents

Specifications

Alternator

Type	Three phase, series wound
Output	12 volt 0.3 Kw at 5,000 rpm
Contact breaker gap	0.012 - 0.016 in (0.3 - 0.4 mm)
Dwell angle	$90^\circ \pm 2.5^\circ$
Ignition timing BTDC	5° @ 900 rpm, $37^\circ \pm 2^\circ$ @ 2,300 - 2,600 rpm

Ignition coil

No. of	Two, twin coils
Make	Toyo Denso

Condenser

Capacity	0.24 MF $\pm$ 10%
Insulating resistance	5 M ohm minimum

Spark plugs

Make	NGK or Nippon Denso
Type	D8ES-L or X24ES
Reach	¾ in (19 mm)
Gap	0.024 - 0.028 in (0.6 - 0.7 mm)
Ballast resistor	3.0 ohms $\pm$ 10%

1 General description

1   The Honda Gold Wing is fitted with a 12 volt electrical system, powered by a three phase alternating current generatore (alternator) The alternator is mounted on a double gear shaft incorporating a shock absorber system, and driven from the rear of the engine by the crankshaft.

2   Ignition source power is fed from the battery to the ignition coil primary windings. When the contact breaker opens, the low tension circuit is interrupted and a high voltage is produced in the ignition coil secondary windings by magnetic induction. Two ignition coils are fitted, one each of which supplies power to the spark plugs of one cylinder bank. The contact breaker assembly is fitted with two points units only, and a single cam. Because of this, power from the coil supplying the cylinder on the compression stroke is also supplied to the other cylinder in that bank, although it is not on the compression stroke, so that both spark plugs fire simultaneously.

3   The contact breaker assembly is fitted to the rear of the left-hand cylinder bank, the contact breaker cam being driven off the camshaft, via an automatic timing unit (ATU). This controls the precise point at which the spark occurs, relative to the engine speed.

4   The alternating current (AC) is passed through a Silicon diode rectifier, where it is converted to direct current (DC) and used to charge the battery. Output of the alternator is controlled to 14 - 15v by a solid state regulator.

2 Alternator: checking the output

1   If the charging performance of the alternator is suspect, it can only be checked satisfactorily with the use of a multimeter test instrument and a separate ammeter and voltmeter. As most owners/riders are unlikely to possess equipment of this type, it is advised that the machine be returned to a Honda service agent for testing.

2   If the required test equipment is available, carry out the check on the alternator in conjunction with tests for the allied components, as described in Chapter 7.

3 Ignition Coil: location and testing

1   Two separate ignition coils are fitted, one which supplies HT current to No. 1 and No. 2 cylinders and the other supplying No. 3 and No. 4 cylinder.
2   Both ignition coils are sealed units incorporating non-detachable H/T leads, and are mounted on brackets bolted to the frame main down tube, within the dummy fuel tank forward of the air filter box. Access can be gained from the left-hand side of the machine.
3   If a weak spark, poor starting or misfiring causes the performance of the coils to be suspected, they should be tested by a Honda service agent or an auto-electrician who will have the appropriate test equipment. A faulty coil must be renewed; it is not possible to effect a satisfactory repair. Some indication of the condition of either coil can be obtained by carrying out the following brief check.
4   Remove the contact breaker assembly cover and disconnect the spark plug supressor caps from the HT leads. Carry the first test out on the left-hand coil. Switch the ignition on and turn the engine over until the left-hand points are closed. Hold No. 1 and No. 2 HT leads so that the uninsulated ends are about 1/8in (3.0mm) away from a suitable part of the engine casing. Using a screwdriver with an insulated handle 'flash' the left-hand points apart. If the spark produced at the HT lead ends in able to jump a gap of from 1/8 - 1/4 in (3.0 - 6.0 mm) it is probable that the coil is in good condition. The same test can be used for the other ignition coil. It is unlikely that both ignition coils will fail simultaneously, consequently if it appears that neither coil is operating, it is more probable that the failure is in some other part of the ignition system.
5   A defective condenser in the contact breaker circuit can give the illusion of defective coils and for this reasons it is advisable to investigate the condition of the condenser before condemning the ignition coils. Refer to Section 6 of this Chapter.

4 Contact breaker: adjustment

1   Remove the contact breaker assembly cover from the rear of the left-hand cylinder head. Note that the cover has a paper gasket to prevent the ingress of water. Remove also all four spark plugs, and the small screw cover from the alternator cover.
2   Rotate the engine as required and examine the condition of the points on both contact breaker assemblies. Provided that the points are in good condition, they can be cleaned in situ, using fine grade emery paper backed by a narrow strip of tin; followed by a clean rag to which methylated spirits or carbon tetrachloride has been applied. If the points faces are blackend and burnt, or badly pitted, it will be necessary to remove them for further attention. See Section 5 of this Chapter.
3   Using a spanner applied to the alternator rotor centre bolt, rotate the engine in the normal direction of running (turn the alternator bolt clockwise), until the left-hand contact breaker points are fully open. Note that the cam has two lobes, one of which is punch marked. Use only the marked lobe when adjusting the points gap. Insert a feeler gauge between the points and check that the gap is within the range 0.012 - 0.16in (0.3 - 0.4mm). If the gap is incorrect, slacken the two fixed point holding screws and move the fixed point closer to or further away from the moving point until the gap is 0.012in (0.3mm). Tighten the two screws and recheck. It is important that the points are in the fully open position or a false reading will be made. Additionally, if gap adjustment is followed by ignition timing and the gap is subsequently adjusted correctly, the ignition timing will be incorrect. Rotate the engine further until the right-hand contact breaker points are fully open. Check and reset as for the left-hand unit.
4   Owner/riders who are equipped with a dwell meter may adjust the points without the need of using a feeler gauge. The correct dwell angle is 90° ± 2.5°.
5   Before replacing the cover and gasket, place a few drops of thin oil on the cam lubricating wick. Do not over lubricate, as excess

oil will eventually find its way onto the contact faces, causing the ignition circuit to malfunction.

5 Contact breaker points: removal, renovation and replacement

1   If the contact breaker points are burned, pitted or badly worn, they should be removed for dressing. If it necessary to remove a substantial amount of material before the face can be restored, the points should be renewed.
2   Remove the points as follows, starting with either of the contact breaker units. Loosen the bolt which retains the low tension lead to the spring post, and detach the lead. Remove the two cross-head screws which retain the fixed point plate, and lift the complete unit from position. The moving contact point can be pulled off the pivot post, after detaching the spring and prising the 'E' clip from the pivot post. Note carefully the relative positions of the various insulating washers, in order to aid correct reassembly.
3   The points should be dressed with an oil stone or fine emery paper. Keep them absolutely square throughout the dressing operation, otherwise they will make angular contact on reassembly and rapidly burn away. If emery paper is used, it should be backed by a flat strip of steel. This will reduce the risk of rounding the edges of the points excessively. Ideally, a slightly domed formation of the points faces will improve bedding-in but must not be taken too far or the contacting surfaces will be badly reduced in area.
4   Check, and if necessary, readjust the contact breaker points when they are in the fully open position, as described in the previous section.

6 Condenser: removal and replacement

1   A condenser (more properly referred to as a capacitor) is included in the contact breaker circuit to prevent arcing across the contact breaker points as they separate. The condenser is connected in parallel with the points and if a fault develops, ignition failure is liable to occur.
2   If the engine proves difficult to start, or misfires when running it is possible that the condenser is at fault. To check, view the contact points when the engine is running. If considerable sparking is apparent and the points are badly blackened and burnt, the condenser can be considered unserviceable. In theory, no sparking should be evident between the points when the engine is running. However, it is usually found in practice that a small amount of irregular sparking will occur. This should be disregarded.
3   The condenser is retained on the battery frame, directly below the rectifier, by two screws, the leads being connected by snap connectors. Removal is therefore straight forward. The specific performance of the condenser can be checked using a capacitance tester that requires a power source of 12 volts. The standard capacity is 0.24 mf ± 10%.

7 Ignition timing: checking and resetting manually

1   The ignition timing should be checked and reset only after contact breaker adjustment has been carried out as described in Section 4 of this Chapter. Ignition timing can be carried out manually as follows or when the engine is running, by using a stroboscope as described in the next Section. Setting the ignition timing manually (static) will give quite acceptable results, providing care is exercised.
2   Remove the flywheel timing mark inspection plug, which is located on the upper rear portion of the crankcase, below the fuel filter. Rotate the engine, using a spanner on the alternator bolt, until the 'F - 1' mark on the flywheel is precisely aligned with the index marks on the viewing orifice. It may be easier to align the flywheel mark with a thin piece of stiff wire placed across the index marks on the orifice. Cut the wire to a suitable length so that it cannot be displaced easily and so fall into the crankcase.

3   When the flywheel is in the correct position as described, the left-hand contact breaker points should be just on the verge of opening. This set of points feeds cylinders Nos. 1 and 2. The exact moment at which the contact breakers open can be ascertained easily if a twelve volt bulb is connected between the moving contact point and an earthing position on the engine. With the ignition switched on, the bulb will illuminate when the points open. If the timing is incorrect, slacken off the two screws which clamp the main stator in place. Rotate the plate clockwise or anticlockwise until the timing bulbs flickers. At this point the ignition timing is correct on cylinders Nos. 1 and 2. Tighten the stator plate screws and re-check.

4   Rotate the engine through 180° (½ turn) until the 'F - 2' timing mark on the flywheel is correctly aligned with the index mark. Check the ignition timing on the right-hand contact breaker unit (cylinders Nos. 3 and 4) as described previously. Note that the adjustment of ignition timing on the right-hand points is made by slackening off the two screws which retain the large half-plate upon which the contact breaker unit rests.

5   Replace the cover through which the flywheel is viewed, before starting the engine.

4.3a Check points gap with CLEAN feeler gauge

4.3b Points retaining screws are also gap adjustment screws

5.2 Contact-breaker assembly dismantled for dressing

7.2a Use length of wire to aid ...

7.2b ... setting flywheel in 'F' position

7.3 Left-hand points timing adjustment screws

7.4 Right-hand points timing adjustment screws

8 Ignition timing: checking with a stroboscope

1   Accurate ignition timing can be simplified by the use of a stroposcopic light attached either to the low tension side of the contact breaker assembly or to an HT lead. This will allow the flywheel periphery to be viewed as the engine is running and the ignition timing so discerned. Stroboscopic timing has the addition advantage in that the timing can be checked over the whole revolution range, giving indications of the correct functioning of the ATU.

2   When viewing the flywheel with the engine running, the inspection orifice should be fitted with a transparent cap, to prevent the exit of engine oil. A special cap with a glas centre is available. (Service tool HC 41335, referred to as the timing cap). The glass is scribed with an index line to aid correct observation.

3   Run the engine at 900 rpm. At this speed, the 'F' mark on the flywheel should line up with the index line. Increase the engine speed to 3000 rpm. At this speed the two separate horizontal lines should line up. The timing is then correct on Nos. 1 and 2 cylinders.

4   Check the timing in a similar manner on Nos. 3 and 4 cylinders, with the stroboscope attached correctly for these cylinders. At 900 rpm the index line should align with the 'F - 2' mark on the flywheel.

5   Adjustments can be made by moving the contact breaker assemblies in the same manner as described for manual ignition timing.

9 Automatic timing unit: examination

1   The automatic timing unit mechanism rarely requires attention, although it is advisable to examine it periodically, when the contact breaker is receiving attention. The unit is retained on the camshaft end by a small bolt and washer, through the centre of the integral contact breaker cam, and can be pulled of the shaft after the contact breaker plate is removed or after the points housing is detached.

2   The unit comprises spring-loaded balance weights, which move outwards against the spring tension as the centrifugal forces increase. The balance weights must move freely on their pivots and be rust free. The tension springs must also be in good condition. Keep the pivots lubricated and make sure the balance weights move easily, without binding.

3   The ATU mechanism is fixed in relation to the camshaft by means of a dowel. In consequence the mechanism cannot be replaced in anything other than the correct position. This ensures accuracy of ignition timing to within close limits, although a check should always be made when reassembly of the contact breaker assemblies is complete.

4   Check also the condition of the points operating cam. If inadequate lubrication of the cam has caused scoring or obvious alteration of the profile, the complete unit will require renewal.

10 Spark plugs: checking and resetting the gaps

1   Four NGK D8ES-L or ND X24ES sparking plugs are fitted as standard to the Honda Gold Wing. Certain operating conditions may indicate a change in spark plug grade, but generally the type recommended by the manufacturer gives best all round service.

2   Check the plugs points gap every 6 months or 6,000 miles. This is the recommended maintenance interval given by the manufacturer, but reducing it somewhat would not be harmful. To reset the gap, bend the outer electrode closer to or further away from the central electrode until the gap is within the range 0.024 − 0.026 in (0.6 − 0.7 mm) as measured with a feeler gauge. The gap is usually set to the lower figure to allow for erosion of the electrodes. Never bend the centre electrode or the insulator will crack, causing engine damage if the particles fall into the cylinder whilst the engine is running.

3   With some experience, the condition of the spark plug electrodes and insulator can be used as a reliable guide to engine operating conditions. See the accompanying diagrams.

4   Always carry a spare spark plug of the recommended grade. In the rare event of plug failure, this will allow immediate replacement and prevent nuisance and the strain on the engine when attempting to continue on three cylinders.

5   Whenever the sparking plugs are removed, take the opportunity to clear out the drain channels which pass from the plug wells to the underside of the cylinder heads. If these become blocked and the machine is used in heavy rain the wells with fill up causing shorting in the suppressor caps and complete ignition failure.

6   If the threads in the cylinder head strip as a result of over tightening the spark plugs, it is possible to reclaim the head by means of a Helical thread insert. This is a cheap and convenient method of replacing the threads; most motorcycle dealers operate a service of this nature at an economic price.

7   make sure the plug insulating caps are a good fit and have their rubber seals. They should be kept clean to prevent leakage and tracking of the H/T current. These caps contain the suppressors that eliminate both radio and TV interference.

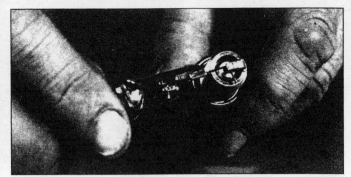

Electrode gap check - use a wire type gauge for best results

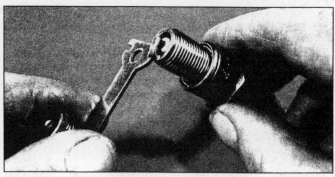

Electrode gap adjustment - bend the side electrode using the correct tool

Normal condition - A brown, tan or grey firing end indicates that the engine is in good condition and that the plug type is correct

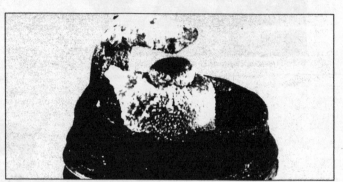

Ash deposits - Light brown deposits encrusted on the electrodes and insulator, leading to misfire and hesitation. Caused by excessive amounts of oil in the combustion chamber or poor quality fuel/oil

Carbon fouling - Dry, black sooty deposits leading to misfire and weak spark. Caused by an over-rich fuel/air mixture, faulty choke operation or blocked air filter

Oil fouling - Wet oily deposits leading to misfire and weak spark. Caused by oil leakage past piston rings or valve guides (4-stroke engine), or excess lubricant (2-stroke engine)

Overheating - A blistered white insulator and glazed electrodes. Caused by ignition system fault, incorrect fuel, or cooling system fault

Worn plug - Worn electrodes will cause poor starting in damp or cold weather and will also waste fuel

9.1a Remove contact breaker housing to gain access ...

9.1b ... to ATU, retained by a single bolt

9.2 Check spring and bob-weight pivot wear

11 Fault diagnosis: ignition system

Symptom	Cause	Remedy
Engine will not start	Faulty ignition switch	Operate switch several times in case contacts are dirty. If lights and other electrics function, switch may need renewal
	Short circuit in wiring	Check whether fuse is intact. Eliminate fault, before switching on again
	Completely discharged battery	If lights do not work, remove battery and recharge
Engine misfires	Faulty capacitor in ignition circuit	Replace capacitor and re-test
	Fouled spark plug	Replace plug and have original cleaned
	Poor spark due to generator failure and discharged battery	Check output from generator. Remove and recharge battery
Engine lacks power and overheats	Retarded ignition timing	Check timing and also contact breaker gap Check whether auto-advance mechanism has jammed
Engine 'fades' when under load	Pre-ignition	Check grade of plugs fitted: use recommended grades only

Chapter 5 Frame and forks

Contents

Specifications

Front forks

Spring length	20.5 in (519 mm)
Service limit	19.5 in (495 mm)
Fork upper tube outside diameter	1.4537 - 1.4547 in (36.925 - 36.950 mm)
Service limit	1.4527 in (36.90 mm)
Slider bush inside diameter	1.4592 - 1.4607 in (37.065 - 37.104 mm)
Service limit	1.4665 in (37.250 mm)

Rear fork

Spring length	9.7873 in (248.6 mm)
Service limit	9.6062 in (244 mm)
Front fork oil capacity:	
After draining	170 - 183 cc (4.7 - 6.1 fl. oz)
After dismantling ... `...	195 - 205 cc (6.6 - 6.9 fl. oz)

1 General description

1 The frame utilised on the Honda Gold Wing is of the full-cradle type; that is, the engine does not comprise any part of the frame. The massive frame incorporates duplex down tubes which run from the steering head lug and either side of the engine, to a point to the rear of the engine, and are additionally strengthened by a tubular cross-member forward of the engine. The frame top tubes and cross tubes are also duplicated, the latter being extended to form a rear frame to which the dualseat and rear mudguard are fixed. The frame down tube on the left of the machine is fitted with a detachable sub-frame in the horizontal section, which can be removed to facilitate engine removal.

The front forks are of the conventional telescopic type, having internal oil-filled dampers. The fork springs are contained within the stanchions (upper tube and lower leg) and each fork leg can be detached from the machine as an individual and complete unit, without dismantling the steering head assembly.

Rear suspension is of the swinging arm type, using oil filled suspension units to provide the necessary damping action. The units are adjustable so that the spring rating can be effectively changed within certain limits, to match the load carried. The right-hand swinging arm member serves also as a torque tube to which the final drive casing is attached, and through which the final drive shaft passes.

2 Front forks: removal

1 It is unlikely that the front forks will have to be removed from the frame s a complete unit, unless the steering head assembly requires attention or if the machine suffers frontal damage.

2 Commence fork removal as a complete unit by detaching the handlebars, disconnecting the controls and switches or removing them completely, if required. Open the left-hand side cover and disconnect the battery before disconnecting any wires, to prevent shorting out and the resultant damage to components. Remove the head lamp reflector unit, complete with rim, so that access may be made to the main wiring connections. The rim is retained by two cross-head screws passing through the shell, on either side. Disconnect all the wires that lead to the handlebar switches, the

instruments and the main frame. Pull the freed wires through the rear of the shell. No difficulty should be encountered when reconnecting wires as they are colour-coded. If there is any doubt, mark the particular wires with tape so that they may be identified later.

3　The hydraulic front brake master cylinder/reservoir can be detached from the handlebar, complete with control leaver, without disconnecting the hydraulic hose. Remove the control clamp, which is retained by two bolts, and lift the complete brake operating assembly away from the bars. Take care not to tip the reservoir so that hydraulic fluid is not spilt. Brake fluid is a very effective paint stripper, and will damage plastics components. Tie the brake cylinder assembly to some part of the machine that is not to be disturbed.

4　Detach the headlamp shell from the mounting 'ears', which extend from the fork shrouds. Unscrew the reflectors and pull off the reflector seats to enable a spanner to be applied to the headlamp retaining bolts.

5　Disconnect the speedometer and tachometer drive cables at the instrument heads by unscrewing the knurled ring on each cable end. Support the instruments and remove the two bolts which retain the instrument mounting bracket to the fork upper yoke. The complete assembly, including the warning lamp console, can be lifted from position. Disconnect the main leads to the ignition switch at the 'block' connector. Remove the ignition switch, after unscrewing the two bolts which pass through the siwtch flange from the underside.

6　Place a sturdy support below the crankcase so that the front wheel is raised well clear of the ground. Remove the two large chrome bolts which retain each brake caliper unit to the fork lower legs. Carefully lift the caliper units from over the discs and tie them to some part of the frame so that they are not supported by the hydraulic hoses. Remove the hose guide clamps from the mudguard, to free the hydraulic hoses. Disconnect the speedometer drive cable at the gearbox on the front hub. The cable is a push fit in the gearbox and is retained by a single screw.

7　Support the front wheel by placing a small wooden block between the tyre and the workshop floor. Loosen and remove the two wheel spindle clamps, which are retained by four nuts each. Remove the block and allow the weel to fall free.

8　Remove the front mudguard, which is retained by one bolt through each stay on the lower end of each fork leg, and by two bolts through brackets on the inside of each lower leg.

9　Unscrew the upper and lower pinch bolts, which pass through the two fork yokes and clamp the fork legs in position. Each fork leg can now be eased downwards, and removed as complete units. It may be necessary to spring the yoke clamps apart with a large screwdriver, to allow the fork legs to leave the yokes. Take care if this method is utilised not to over strain the clamps as the material is quite brittle.

10　Prise the rubber cap from the top of the steering stem. Using a 'C' spanner, loosen and remove the crown nut and washer. Loosen the pinch bolt which passes horizontally through the rear of the upper (crown) yoke. With the aid of a rawhide mallet, the upper yoke can be tapped upwards and off the steering stem. At the same time as the yoke is moved upwards, the two fork shrouds, complete with the flashing indicators and shroud guides, can be removed.

11　Unbolt the hydraulic hose two-way joint from the lower yoke. Detach the previously tied brake components and lift the interconnected components away from the machine as one assembly.

12　To release the lower yoke and the steering head stem, unscrew the adjuster ring at the top with a suitable 'C' spanner. If such a spanner is not available, a soft brass drift and hammer may be used to loosen the ring. As the yoke and stem are lowered, the uncaged ball bearings from the lower race will be released, and care should be taken to catch them as they fall free. The bearings in the upper race will almost certainly stay in place.

13　It follows that much of the procedure can be avoided if it is only necessary to remove the individual fork legs, without disturbing the fork yokes and steering head bearings. Under these circumstances commence dismantling at paragraph 6 and work through to paragraph 9.

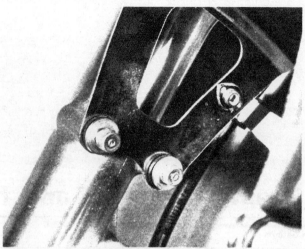
2.8a Mudguard is retained by bolts into fork leg ...

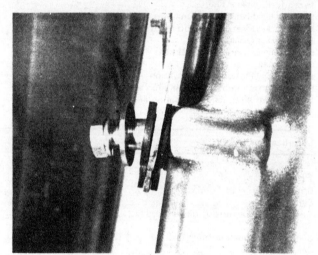
2.8b ... note the rubber cushion and spacer

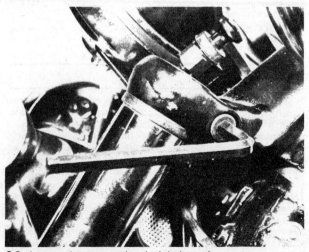
2.9a Loosen the upper yoke pinch bolts and ...

2.9b bolts through lower yoke to allow ...

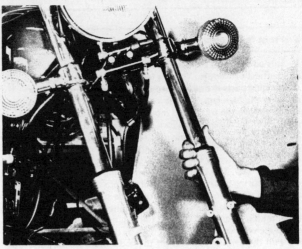

2.9c ... removal of each leg separately

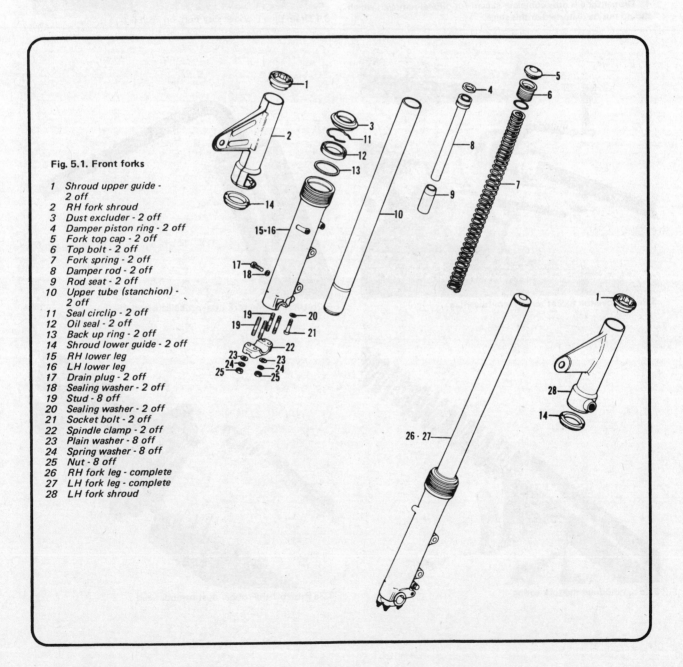

Fig. 5.1. Front forks

1   Shroud upper guide -
    2 off
2   RH fork shroud
3   Dust excluder - 2 off
4   Damper piston ring - 2 off
5   Fork top cap - 2 off
6   Top bolt - 2 off
7   Fork spring - 2 off
8   Damper rod - 2 off
9   Rod seat - 2 off
10 Upper tube (stanchion) -
    2 off
11 Seal circlip - 2 off
12 Oil seal - 2 off
13 Back up ring - 2 off
14 Shroud lower guide - 2 off
15 RH lower leg
16 LH lower leg
17 Drain plug - 2 off
18 Sealing washer - 2 off
19 Stud - 8 off
20 Sealing washer - 2 off
21 Socket bolt - 2 off
22 Spindle clamp - 2 off
23 Plain washer - 8 off
24 Spring washer - 8 off
25 Nut - 8 off
26 RH fork leg - complete
27 LH fork leg - complete
28 LH fork shroud

3 Front forks: dismantling

1 It is advisable to dismantle each fork leg individually, using an identical procedure. There is less chance of unwittingly exchanging parts if this approach is adopted. Commence by draining the fork legs; there is a drain plug in each lower leg, located above the mudguard mounting stay bolt. The draining rate can be increased if the fork legs are worked up and down.

2 Remove the socket screw from the extreme lower end of the lower fork leg. This screw retains the damper rod to the bottom of the fork. Unscrew the recessed fork top bolt, after prising off the rubber cap. The bolt is unscrewed using an Allen key. Pull the fork spring out of the upper tube. Note that the close coiled section of the spring faces upwards in the tube.

3 Carefully prise the rubber dust excluder from around the upper tube, taking care not to damage the sealing lip. The upper tube, complete with damper rod, can now be drawn out of the lower leg. Pull the damper rod seat off the rod and invert the fork tube. The damper rod will fall out.

4 Dismantling is now complete except for oil seal removal, which should not be disturbed at this stage.

3.1 Drain plug is above stay bolt, on each leg

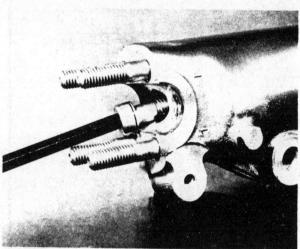

3.2a Remove the socket screw from lower leg

3.2b Unscrew the fork upper tube caps and ...

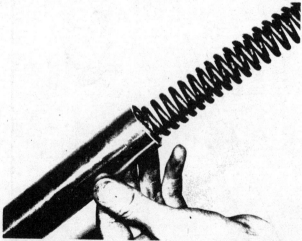

3.2c ... withdraw the fork spring

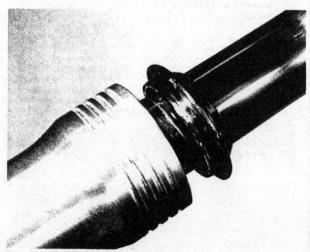

3.3a Prise off the rubber dust excluder and ...

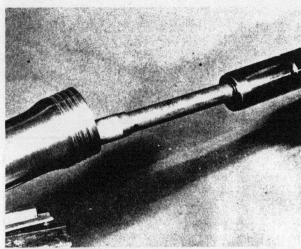

3.3b ... Pull complete upper tube fork leg

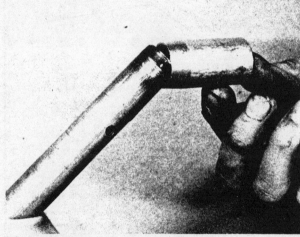

3.3c Remove the push fit damper rod seat and ...

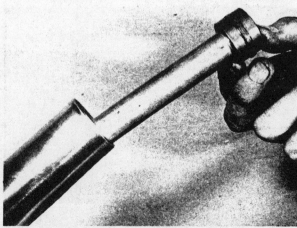

3.3d ... allow the damper rod to fall out

4 Steering head bearings: examination and renovation

1 Clean and examine the cups and cones of the steering head bearings. They should have a polished appearance and show no signs of indentation. Renew the set if necessary.
2 Clean and examine the ball bearings which should also be polished and show no signs of surface cracks or blemishes. If any require replacement the whole set must be renewed.
3 All the balls are 8 mm diameter (do not mix metric and English sizes as they are slightly different). Eighteen balls are fitted in the top race and nineteen in the bottom. This arrangement will leave a gap but an extra ball must not be fitted otherwise the balls will press against each other, accelerating wear and making the steering stiff.

5 Front forks: examination and renovation

1 The parts most likely to wear over an extended period of service are the internal surfaces of the lower leg and the outer surfaces of the fork stanchion or upper tube. Check the upper tube for scoring over the length which enters the oil seal. Bad scoring here will lead to a damaged seal and leakage of damping fluid.
2 Measure the outside diameter of the upper tube on the wearing length and the internal diameter of the lower leg. If wear exceeds that given for the service limit, the fork components should be renewed. It is advisable to renew the upper tube and lower leg at the same time, even if only one component has worn excessively.

Front fork upper tube	
Outside diameter	1.4537 - 1.4547 in
	(36.925 - 36.950 mm)
Service limit	1.4527 in
	(36.900 mm)
Slider bush	
Inside diameter	1.4592 - 1.4607in
	(37.065 - 37.104 mm)
Service limit	1.4665 in
	(37.250 mm)

The slider bush is integral with the fork lower leg and can not be removed for renewal.
3 It is advisable to renew the oil seals when the forks are dismantled even if they appear to be in good condition. This will save a strip-down of the forks at a later date if oil leakage occurs. The oil seal in the top of each fork lower leg is retained by an internal 'C' ring which can be prised out of position using a small screwdriver. Do not remove the oil seals if they need not be renewed. Removal will invariably distort the seal or tear the sealing lip, rendering the seal useless for further service.
4 It is not generally possible to straighten forks which have been badly damaged in an accident, particularly when the correct jigs are not available. It is always best to be on the safe side and fit new ones; especially since there is no easy way of detecting whether the forks have been overstressed and suffered metal fatigue. The upper tubes (stanchions) can be checked, after removal from the lower legs, by rolling them on a dead flat surface. Any misalignment will be immediately obvious.
5 The fork springs will take a permanent set after considerable service and will need renewal if the fork action become spongy. The service limit for the total free length of each spring is 19.5 in (495 mm). The springs should always be renewed as a matched pair.

6 Front fork: replacement

1 Replace the front forks by reversing the dismantling procedures described in Section 2 and 3 of this Chapter. Apply a sealing fluid

to the socket screws, before replacing them in the lower legs. The close coiled section of each fork spring must be fitted uppermost in the fork tubes. Before fully tightening the front wheel spindle clamps and the fork yoke pinch bolts, bounce the forks several times to ensure they work freely and are clamped in their original positions. Complete the final tightening from the front wheel spindle upwards.

2   Do not forget to add the damping fluid to each fork leg. The correct quantity after complete dismantling is 195   205 cc (6.6 - 6.9 fl. oz) of good quality AFT (Automatic transmission fluid) or Fork oil. The oil can be replaced most easily when the legs are not fitted to the machine. In any event, the filling must take place before replacement of the handlebars. Check that the drain plugs have been fitted and tightened, before filling.

3   If the fork upper tubes prove difficult to re-locate through the fork yokes, make sure their outer surfaces are clean and polished so that they will slide more easily.

4   Before the machine is used on the road, check the adjustment of the steering head bearings, if they are too slack, judder will occur. Refer to the next Section for the adjustment procedure.

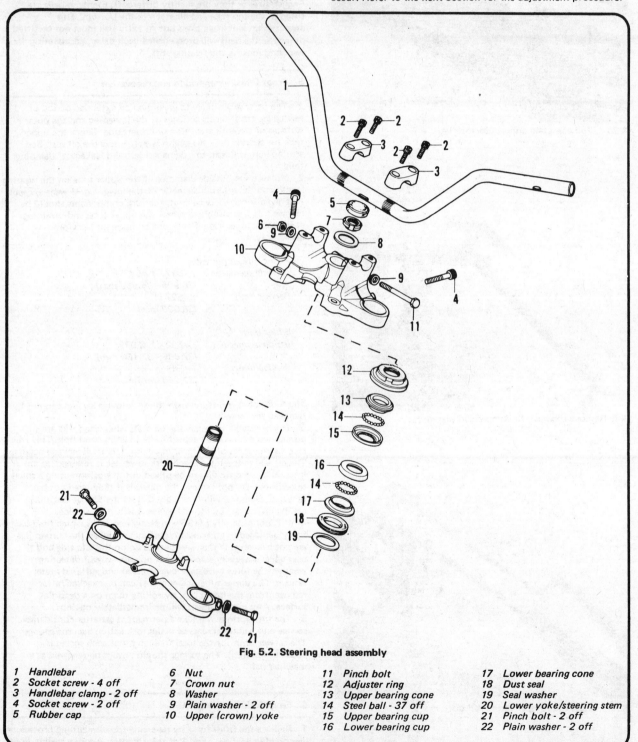

Fig. 5.2. Steering head assembly

1	Handlebar	6	Nut	11	Pinch bolt	17	Lower bearing cone
2	Socket screw - 4 off	7	Crown nut	12	Adjuster ring	18	Dust seal
3	Handlebar clamp - 2 off	8	Washer	13	Upper bearing cone	19	Seal washer
4	Socket screw - 2 off	9	Plain washer - 2 off	14	Steel ball - 37 off	20	Lower yoke/steering stem
5	Rubber cap	10	Upper (crown) yoke	15	Upper bearing cup	21	Pinch bolt - 2 off
				16	Lower bearing cup	22	Plain washer - 2 off

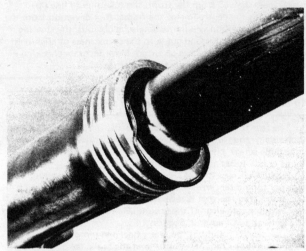

5.3a Internal circlip in fork lower leg ...

5.3b ... retains oilseal which should be examined carefully

5.3c Check damper piston and ring for scoring

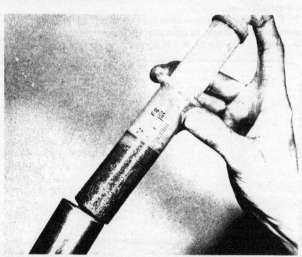

6.2a Refill each fork leg with correct quantity of fluid

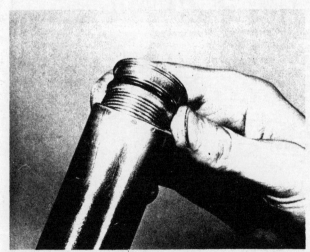

6.2b Do not omit 'O' ring on fork tube cap

7 Steering head bearings: adjustment

1   The steering head bearings should be adjusted so that the forks turn freely from side to side, with no resistance, If, with the bike on the centre stand and the front wheel in line with the frame, the handlebar is pushed, the forks should fall easily to one side. At the same time, there should be no free play in the bearings. Maladjusted steering head bearings are a frequent cause of handling problems. If the forks seem to 'index' in one position when turned, check the bearing tracks for pitting. If the bearings are too tight. the machine will have a characteristic 'roll' at low speeds.

2   With the machine on its centre stand, grasp the forks on each side, at the bottom. Pull and push on the forks horizontally. No movement should be discernable. If another person is able to help, any movement can be felt by putting ones fingers between the top yoke and the steering head.

3   If adjustment is required slacken the pinch bolt that passes through the rear of the upper fork yoke, and also the crown nut. Using a 'C' spanner, tighten the adjuster nut below the upper yoke until the bearings are free from play. Re-tighten the pinch bolt and crown nut and recheck the adjustment.

8 Steering head lock:

1   On some models the steering lock is fitted within a lug on the left-hand side of the steering head lug, and is retained by a Mills pin. If the lock fails, or the keys are lost, the lock cylinder can be removed after drifting out the pin with a punch.

2   The 'General Type' model incorporates a steering lock in the ignition switch, which is located in the warning lamp console on the handlebars. The forks become immovable when the key is turned to the 'lock' position and removed.

9 Frame: examination and renovation

1   If the machine is stripped for an overhaul, this affords an excellent opportunity to inspect the frame for signs of cracks or other damage which may have occurred during service. Frame repairs are best entrusted to a frame repair specialist who will have all the necessary jigs and mandrels necessary to ensure correct alignment. This type of approach is recommended for minor repairs. If the machine has been damaged badly as the result of an accident and the frame is well out of alignment, it is advisable to renew the frame without question or, if the amount of money available is limited, to obtain a sound replacement from a breaker's hard.

2   If the front forks have been removed from the machine, it is comparatively simple to make a quick visual check of alignment by inserting a long tube that is a good push fit in the steering head races. Viewed from the front, the tube should line up exactly with the centre line of the frame. Any deviation from the true vertical position will immediately be obvious; the steering head is a particularly good guide to the correctness of alignment when front end damage has occurred. More accurate checking must be carried out when the frame is stripped.

10 Swinging arm: removal, examination and renovation

1   The rear fork assembly pivots on needle roller bearings, fitted each side of the fork tubular cross member, which bear on adjustable screw pivot stubs fitted into the lugs, forward of the frame cradle tubes and downtubes.

2   When wear necessitates bearing renewal, rear suspension removal should be carried out as follows: Place the machine on the centre stand so that it rests securely on level ground and the rear wheel is well clear of the ground. Remove the exhaust-system as described in Chapter 3, Section 14.

3   Support the weight of the rear wheel and remove the rear suspension lower mounting bolts, and the brake caliper stay bolt. Remove the split pin from the rear wheel spindle nut and unscrew the nut. Draw the wheel spindle out towards the left-hand side of the machine. Lift the brake caliper from over the brake disc and away from the machine, after pulling the wheel spacer from position. Pull the rear wheel across so that the splined drive disengages from the hub centre. The rear wheel can now be lifted from the machine.

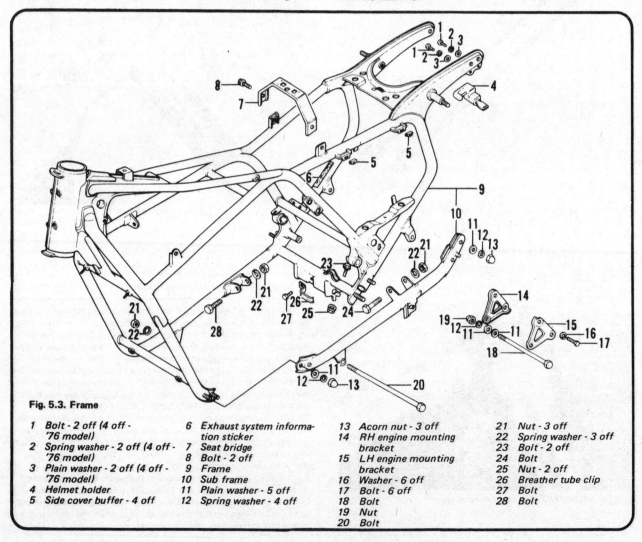

Fig. 5.3. Frame

1 Bolt - 2 off (4 off - '76 model)	6 Exhaust system informa- tion sticker	13 Acorn nut - 3 off
2 Spring washer - 2 off (4 off - '76 model)	7 Seat bridge	14 RH engine mounting bracket
3 Plain washer - 2 off (4 off - '76 model)	8 Bolt - 2 off	15 LH engine mounting bracket
4 Helmet holder	9 Frame	16 Washer - 6 off
5 Side cover buffer - 4 off	10 Sub frame	17 Bolt - 6 off
	11 Plain washer - 5 off	18 Bolt
	12 Spring washer - 4 off	19 Nut
		20 Bolt

21   Nut - 3 off
22   Spring washer - 3 off
23   Bolt - 2 off
24   Bolt
25   Nut - 2 off
26   Breather tube clip
27   Bolt
28   Bolt

4 Temporarily replace the left-hand suspension unit lower bolt so that the weight of the gear housing is taken. Loosen and remove the three nuts which retain the final drive housing to the torque tube flange. Lift the housing away.

5 Remove the external circlip from the inside of the splined joint at the output end of the final drive shaft. The circlip is deeply recessed so will require a pair of long jawed circlip pliers for easy removal. Pull the splined joint off the shaft. Prise the rubber boot of the torque tube/ engine casing joint and remove the external circlip that retains the U-joint to the splined output shaft. Pull the joint off the splines.

6 Free the rear brake hydraulic hose from the hose clip on left-hand fork member and prise the rubber caps off the swinging arm pivot locknuts. Loosen the locknuts and slowly unscrew the pivots, using a suitable Allen key. Take the weight of the swinging arm assembly as the pivot stubs leave the bearings. Lift the swinging arm fork out towards the rear of the machine, tilting it slightly to avoid knocking the rear brake master cylinder. Pull the final drive shaft out of the torque tube from the forward end.

7 Remove the bearing covers from either side of the fork cross member and pull out the 'Tophat' shaped inner bearing pieces. The needle roller bearings are retained in blind outer casings which are capped by a shouldered insert. Drift out the right-hand bearing, after prising the shouldered insert from position. The inserts are manufactured from a brittle material and will probably fracture during removal. This is difficult to avoid. Using a long drift, knock out the left-hand bearing and insert from the right-hand side.

8 The needle roller bearings and the shouldered inserts should not be disturbed unless they are to be renewed. As mentioned, the inserts will probably fracture and the bearing cases will probably become distorted.

9 New bearings should be drifted into position from the outside using a close fitting drift so as not to damage the cases or cages. Each bearing must be inserted sufficiently, so that shouldered portion of the inserts is exactly flush with the outer face of the cross member. This is important so that the cylindrical portion of the inserts are not subjected to excessive compression when in service.

10 Refit the swinging arm assembly by reversing the dismantling procedure and by following the directions that follow. Grease the swinging arm bearings and the drive shaft splines thoroughly with a lithium based grease that has a molybdenum disulphide additive.

11 When replacing the final drive housing, refit the wheel spindle temporarily so that the housing is correctly aligned, before tightening the flange nuts.

12 The swinging arm pivots must be adjusted as follows if the rear suspension is to function efficiently. This adjustment should take place with the final drive housing refitted but without the rear wheel and without reattaching the suspension units. Screw both pivot stubs inwards until about two (2) thread pitches extend from the outer face of the locknuts. Screw the right-hand pivot inwards and tighten the locknut fully so that one (1) thread pitch extends from the outer face of the locknut. Torque tighten the left-hand pivot stub to 7 lb. ft (100 kg - cm) and check that the swinging arm moves up and down smoothly. If the movement is tight, loosen the pivot stub through a range of 0° - 60° until correctly adjusted. There must be no side movement when the adjustment is correct. Tighten the locknut and recheck. Worn swinging arm bearings will cause handling problems, making the rear end twitch and hop, especially when accelerating or shutting off whilst banked over. Play in the bearings may be detected by pulling and pushing horizontally on the rear fork ends. Any play will be magnified by the leverage obtained.

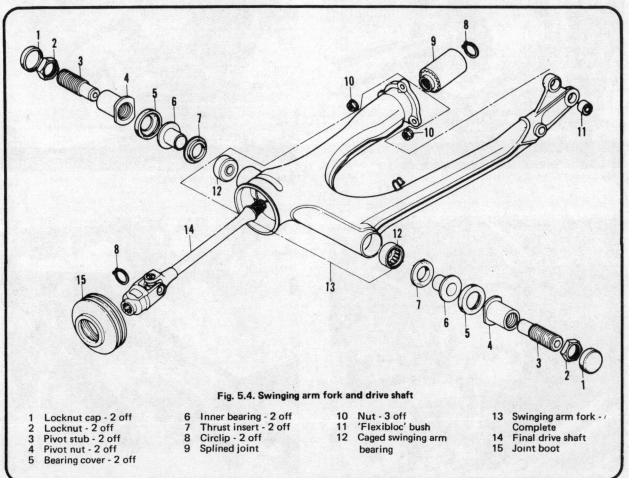

Fig. 5.4. Swinging arm fork and drive shaft

1	Locknut cap - 2 off	6	Inner bearing - 2 off	10	Nut - 3 off	13	Swinging arm fork -
2	Locknut - 2 off	7	Thrust insert - 2 off	11	'Flexibloc' bush		Complete
3	Pivot stub - 2 off	8	Circlip - 2 off	12	Caged swinging arm	14	Final drive shaft
4	Pivot nut - 2 off	9	Splined joint		bearing	15	Joint boot
5	Bearing cover - 2 off						

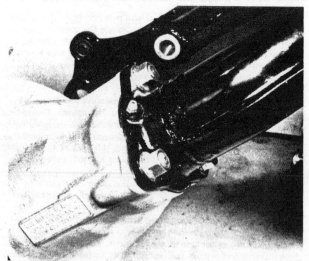

10.4a Remove the three retaining bolts and ...

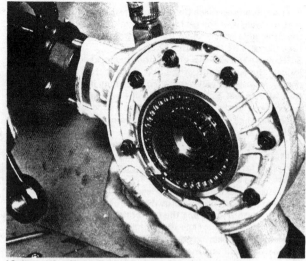

10.4b ... lift off the final drive gear housing

10.5a Extract external circlip ...

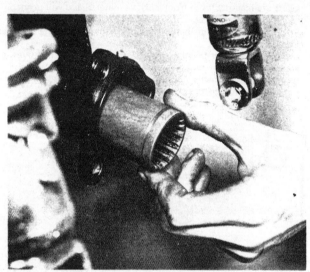

10.5b ... pull off splined drive joint

10.6a Remove locknuts and unscrew pivot stubs to ...

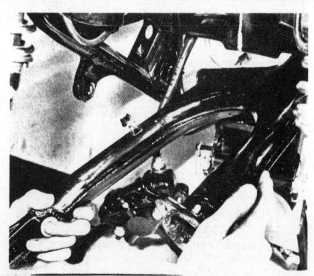

10.6b ... enable swinging arm fork to be lifted out

10.7a Remove bearing cover followed by ...

10.7b ... the 'top hat' inner bearing races

10.7c Shouldered thrust insert will break on removal

10.7d Bearings are an interference fit in swinging arm

10.7e Pivot threads are in detachable housing

11 Rear suspension units: examination and renovation

1   The units are adjustable for load in five positions. Turn the adjuster clockwise with the toolkit, C spanner to increase the spring preload. Both units must be at the same setting.

2   Examine the damper units for oil leaks. Bounce the rear of the machine. If it does not come to rest after a couple of oscillations, the dampers may be faulty.

3   The dampers are sealed units, and cannot be repaired. If oil leaks are apparent, the oil seal is damaged and the damper requires renewal. It is best to renew both damper units, as a matched pair.

4   To remove the damper unit, it is necessary first to take off the lifting handle. This is fixed to the suspension unit top mounting, and to the rearward frame extensions. Unscrew the suspension unit lower mounting bolt, and pull off the unit.

5   Hold the lower mounting in a vice and depress the spring cover until the split cotters can be removed. It is best to employ a second person to remove the collets whilst the spring is compressed. Remove the cover and spring. It should be possible to feel an equal resistance throughout the travel of the damper rod, in both directions. Check the spring free length.

6   Reverse this procedure to reassemble the suspension unit.

12 Centre stand: examination

1   The centre stand pivots on a hollow tube, clamped between two split lugs below the rear engine mounting. The pinch bolts should be checked occasionally, for security and the pivot oiled.
2   Also check that the return spring and linkage is unworn and retracts the stand smartly. If the stand falls whilst the bike is in motion, an accident may result.

13 Prop stand: examination

1   The prop stand is pivoted on a lug under the front left-hand engine mount. Check the security of the pivot bolt and oil occasionally.
2   Check the action of the stand, and make sure that the return spring is not worn or weakened. An accident will almost inevitably result from the prop stand dropping whilst the machine is in motion.
3   To prevent the machine from being ridden away with the prop stand down, American models incorporate a novel self-retracting device. Check the rubber 'trip' of this device for wear or damage. No part should be worn below the moulded line on the rubber.
4   Check the operation of the stand as follows: Put the machine on the centre stand, and put the prop stand down. Using a spring balance attached to the extreme end of the centre stand, measure the force required to retract the stand. If this force exceeds 2 - 3 kg (4.4 - 6.6 lb), check that the stand pivot bolt is not overtight, or in need of lubrication.
5   To renew the rubber 'trip', take off the bolt. Make sure the sleeve is installed in the fixing hole of the new trip. Fit the trip with the arrow facing outwards. The block should be marked 'over 260 lbs only'.

14 Footrests: examination

1   The front footrests are retained on the frame by a single bolt and comprise a spring loaded rubber covered peg attached by a clevis pin to a forged arm. The footrest rubber is held to the support arm by two bolts passing through a plate below the rubber.
2   Because the footrest pegs are spring loaded and can fold back when struck, it is unlikely that they will become damaged. The support arms can be repaired, if bent, by the application of heat before being straightened.
3   To renew a returnspring, remove the split pin and pull out the clevis pin from the footrest pivot piece.
4   The pillion footrests are also of the folding type, spring loading being provided by the compression of the footrest rubber. Owing to the construction, it is unlikely that these footrests can be repaired, if damaged.

15 Rear brake pedal: examination

1   The rear brake pedal is mounted on, and pivots on, a shaft fixed inboard of the frame, just below the swinging arm pivot stub on the right-hand side of the machine. The pedal is retained by a washer and split pin. The pedal return spring is fitted on the pivot shaft between the pedal and washer.
2   The brake pedal is connected directly to the piston rod of rear brake master cylinder by means of a forked joint and clevis pin. The piston rod and fork are threaded, to allow height adjustment of the pedal.
3   If the pedal becomes bent during an accident, it may be straightened, after removal from the machine. Apply heat to the damaged area with a blowtorch or gas bottles. Bear in mind that the required amount of heat will probably cause the chrome plating to peel from the pedal. This is unavoidable.

16 Dualseat: removing

1   The dualseat is retained at the rear by two bolts passing through brackets attached to the seat pan. The seat front is supported on a rubber saddle mounted on a pressed steel bridge across the frame tubes.
2   The seat can be removed with ease, after removal of the two bolts.

17 Dummy fuel tank: examination

1   The dummy fuel tank, fitted in place of a receptacle for fuel, consists of three fibre glass panels containing the air filter box, an electrical components board and the radiator reservoir.
2   The upper panel is fitted with a lockable catch and houses the fuel gauge. Opening the upper panel gives access to the screw knobs, which allow the side panels to open and hinge about their lower edges. The side panels are prevented from opening beyond a certain position by cable straps fitted with barrel nipples at each end.

18 Instrument drive cables: examination and replacing

1   Drive cables should be examined and lubricated occasionally. The outer sheath should be examined for cracks or damage, the inner cable for broken or frayed strands. Jerky or sluggish instrument movement is generally caused by a faulty cable.
2   Detach the cable at the drive end, and withdraw the inner cable. Clean and examine the cable. Re-lubricate it with high melting point grease, but do not grease the top six inches of cable, at the instrument end, or grease will work its way into the instrument head and ruin the movement.
3   Re-route the cables as they were originally. Make sure that the steering turns freely.

19 Instrument heads: removing

1   Unscrew the knurled cable nut and disconnect the drive cable. Unscrew the two acorn nuts and remove the large washer, grommet and sleeve. Lift the head and pull out the bulb holders. Remove the head.
2   It is not possible to repair a faulty instrument head. If the instrument fails completely, or moves jerkily, first check the drive cable. If the mileage recorder of the speedometer ceases to function but the speedometer continues to work or vice versa, the instrument head is faulty.
3   Remember that a working speedometer, accurate at 30 mph, is a statutory requirement in the UK and many other countries.

20 Cleaning the machine

1   If possible, the machine should be wiped down immediately after use in the wet, so that it is not garaged in a rust-promoting condition.
2   Before regular cleaning, wash off dirt with plenty of water, and allow the machine to dry. Use a wax polish on painted parts, and a proprietary chrome cleaner such as Solvol Autosol, on plated and aluminium alloy items. Note that some alloy parts on these motorcycles have a lacquered finish which needs wiping only.
3   If any part of the machine is caked with an oily film, use a cleaner such as Gunk, in accordance with the instructions. Keep water out of the carburettor and ignition components. After washing down the machine, re-oil the exposed control cables, and bearings.

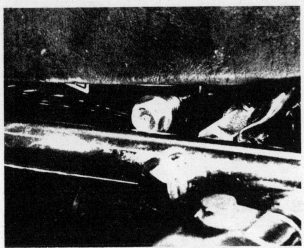

16.1a Dualseat is retained by two bolts at rear and ...

16.1b ... supported by rubber saddle on bridge at front

21 Fault diagnosis: frame and forks

Symptom	Cause	Remedy
Machine is unduly sensitive to road conditions	Forks and/or rear suspension units have defective damping	Check oil level in front forks. Renew rear suspension units
Machine tends to roll at low speeds	Steering head bearings overtight or damaged	Slacken bearing adjustment. If no improvement, dismantle and inspect bearings
Machine tends to wander, steering is imprecise	Worn swinging arm bearings	Check and if necessary renew bearings
Fork action stiff	Fork legs have twisted in yokes or have been drawn together at lower ends	Slacken off spindle nut clamps, pinch bolts in fork yokes and fork top nuts. Pump forks several times before retightening from bottom
Forks judder when front brake is applied	Worn fork bushes Steering head bearings too slack	Strip forks and renew bushes Re-adjust to take up play
Wheels out of alignment	Frame distorted as result of accident damage	Check frame alignment after stripping out. If bent, specialist repair is necessary

Chapter 6 Wheels, brakes and final drive

Contents

Specifications

Tyres

Front	3.50 H19 (19 inch diameter)
Rear	4.50 H17 (17 inch diameter)
Tyre pressures solo (cold)	
Front	28 psi (2.0 kg/sq. cm)
Rear	32 psi (2.25 kg/sq. cm)
Tyre pressures two-up (cold)	
Front	28 psi (2.0 kg/sq. cm)
Rear	40 psi (2.8 kg/sq. cm)

Brakes

Front	Hydraulically operated twin disc
Rear	Hydraulically operated single disc

---

1 General description

Both wheels are fitted with steel rims, laced to alloy hubs, which
are supported on two ball journal bearings. The front wheel is
fitted with a twin disc hydraulic brake, the calipers being of the
self-aligning type, using a single piston and located to the rear of
the fork lower legs. The rear brake is also hydraulically operated,
with a single disc attached to the left-hand side of the hub. The
caliper used is of the fixed type, where two pistons are used, each
one applying pressure to a disc pad. The front brake master
cylinder and reservoir is mounted in an exposed position on the
handlebar, where it is operated directly by the lever. The rear
master cylinder and reservoir is mounted inboard of the frame,
to the rear of the brake pedal, and is operated by a rod joined
by fork and pin to the trailing edge of the brake pedal. Final
drive is provided by a shaft running through the right-hand
tubular member of the swinging arm, to which is attached a
crown and pinion final drive assembly contained within an alloy
housing. A cush drive assembly is interposed between the final
drive gear and the rear wheel, to absorb shocks in the transmission.

2 Front wheel: examination

1   Place the machine on its centre stand, with a block under the
sump, so that the front wheel is clear of the ground.
2   Spin the wheel and check for rim alignment. Small
irregularities can be corrected by tightening the spokes in the
affected area, although a certain amount of experience is
necessary if over-correction is to be avoided. Any 'flats' in the
wheel rim should be evident at the same time. These are more
difficult to remove with any success, and in most cases the wheel
will have to be rebuilt on a new rim. Apart from the effect on
stability, especially at high speeds, there is much greater risk of
damage to the tyre beads and walls if the machine is ridden with
a deformed wheel.
3   Check for loose or broken spokes. Tapping the spokes is the
best guide to tension. A loose spoke will produce a quite
different note and should be tightened by turning the nipple in
an anticlockwise direction. Always check for run-out by spinning
the wheel again. If a spoke has been tightened, the tyre should
be removed to ensure that the spokes do not protrude beyond the

nipple, when they may puncture the tube. File or grind off the protruding ends.

4   Grasp the wheel at its periphery, and push and pull on the rim, to check for play in the wheel bearings.

3   Front wheel: removal and replacement

1   With the front wheel supported well clear of the ground, remove the cross-head screw which retains the speedometer cable to its drive gearbox, on the right-hand side of the hub. Pull the cable out and replace the screw, to prevent loss.

2   Remove both brake caliper units, tying them to the mudguard stays so that the weight of each caliper is not taken by the hydraulic hose. Each caliper is retained by two large chrome bolts, which pass into lugs on the fork lower leg.

3   Place a block of wood under the front tyre to take the weight of the wheel. Undo the four nuts on each wheel spindle clamp and remove the nuts, washers and the two clamps. Take the weight of the wheel, remove the block, and lower the wheel from position.

4   Do not operate the front brake lever while the wheel is removed since fluid pressure may displace the pistons and cause leakage. Additionally, the distance between the pads would be reduced, making refitting of the brake discs more difficult.

5   To refit the wheel, place the wheel between the forks. Lift the wheel upwards so that it locates with the spindle caps in the fork legs and place the wooden block under the tyre. Make sure that the lug on the speedometer drive gearbox rests against the rear of the lug on the right-hand fork leg. Install the spindle clamps with the 'F' marks facing forwards and tighten the forward nuts lightly. Replace the brake calipers, taking care that the disc enters the space between the pads squarely so that the pads are not chipped. Tighten the caliper bolts fully.

6   Tighten the right- hand spindle clamp nuts down evenly to a torque of 156 - 216 in. lb (180 - 250 kg. cm). Measure the clearance between the brake disc and the outside rear of the left-hand caliper unit with a feeler gauge. If the clearance is correct 0.028 in (0.7 mm) tighten the left-hand spindle clamp to the specified torque setting. If the clearance is incorrect, pull the left-hand fork leg outwards to adjust and then tighten the clamp nuts.

7   Spin the wheel to ensure that it revolves freely and check the brake operation. Check that all nuts and bolts are full tightened. If the clearance between the discs and pads is incorrect, pump the front brake lever several times to adjust.

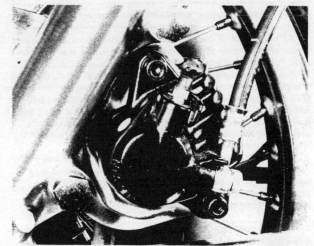

3.2 Brake calipers are retained by two large bolts

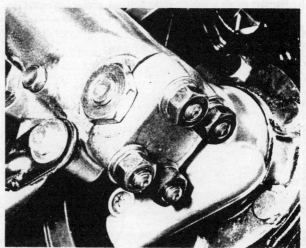

3.3 Each spindle clamp is retained by four nuts

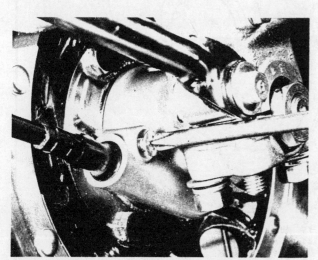

3.1 Speedometer cable is retained by single screw

3.5 Lug on gearbox must rest against lug on fork

4   Front wheel bearings: examination and renovation

1   After removal of the front wheel, place a spanner on each end of the wheel spindle and unscrew the sleeve nut on the right-hand side. Pull the spindle out of the hub centre.

2   Six long bolts pass through the right-hand brake disc, through the gearbox drive retainer and through the hub and left hand brake disc. Remove the self-locking nuts, push out the bolts and pull the left-hand brake disc off the boss on the side of the hub. Pull the speedometer gearbox from position followed by the packing piece and speedometer drive dog. Note the 'O' ring. Then pull the right-hand brake disc off the boss.

3   Unscrew the bearing retaining ring on the left-hand side of the hub. This is of aluminium alloy, staked into the hub, and very easily damaged by punching round. The grease seal also is pressed into this ring and will be damaged too. Use only a properly fitting tubular tool, unless both items are to be renewed.

4   The left-hand side bearing may be driven out from the opposite side using a soft metal drift which locates on the inner diameter of the bearing spacer. When this bearing and the spacer have been removed, the right hand bearing can be driven out, using the same drift.

5   Remove excess grease from the bearings, spacer and inside the hub. Clean the bearing housings in the hub. Wash the bearings in white spirit to remove all grease. Do not spin a dry bearing. If the bearings show more than very slight radial play, or roughness when rotated, they should be renewed.

6   A scorched, glazed appearance on the bearing housing and the outer diameter of the bearing will indicate that it has been rotating in the hub. If damage to the hub is not too great, replace the bearing using Loctite or a similar bearing adhesive to effect a cure.

7   Pack the bearings with grease before refitting. Do not over grease, as this will only raise the running temperature of the bearing, and create drag. Tap the bearings into the hub, not omitting the spacer. Tap only the outer ring of a bearing never the inner. A tube is the best tool to use, ensuring that the bearings enter the housings squarely. It is an advantage to freeze the bearings before fitting, to decrease the interference fit. The integral seals must face outwards.

8   Check the 'O' ring seal on the hub spigot, and replace all parts in reverse order. Check the condition of the multi-lip seal in the retaining ring and renew it if it is scratched or damaged. Use Loctite or any similar adhesive to lock the ring in position, without staking. The speedometer gearbox should be engaged with the two drive tongues on the drive piece. Lift the wheel spindle from the left-hand side after lightly greasing it with a multi-purpose grease.

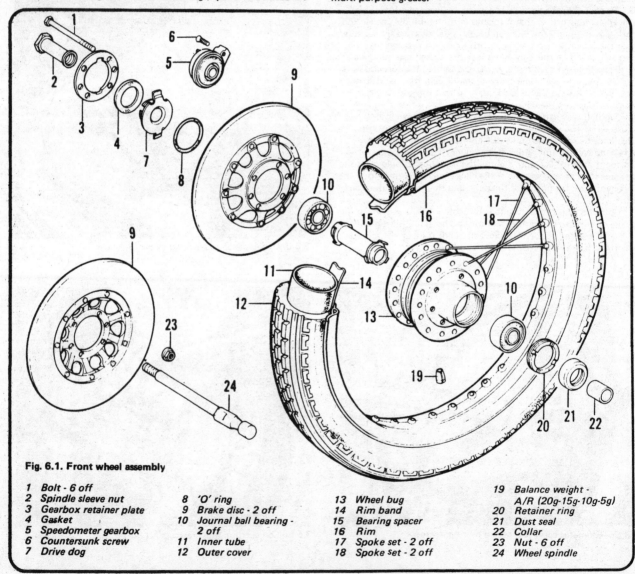

Fig. 6.1. Front wheel assembly

1 Bolt - 6 off		
2 Spindle sleeve nut	8 'O' ring	13 Wheel bug
3 Gearbox retainer plate	9 Brake disc - 2 off	14 Rim band
4 Gasket	10 Journal ball bearing -	15 Bearing spacer
5 Speedometer gearbox	2 off	16 Rim
6 Countersunk screw	11 Inner tube	17 Spoke set - 2 off
7 Drive dog	12 Outer cover	18 Spoke set - 2 off

19   Balance weight -
A/R (20g-15g-10g-5g)
20   Retainer ring
21   Dust seal
22   Collar
23   Nut - 6 off
24   Wheel spindle

4.1 Remove nut and withdraw spindle

4.2a Pull left hand disc off hub boss

4.2b Withdraw the speedometer gearbox followed by ...

4.2c ... the drive dog retainer plate and ...

4.2d ... the drive dog itself

4.3a Drift out wheel bearings and ...

4.3b ... remove bearing spacer from hub

4.9 Note gasket below speedometer dog plate

6.1 Remove rear suspension unit lower bolts

5 Rear wheel: examination

1   Place the machine on its centre stand, with the rear wheel clear of the ground.
2   Spin the wheel and make the checks as described for the front wheel, in Section 2 of this Chapter.

6 Rear wheel: removal and replacement

1   Place the machine on the centre stand and remove the rear suspension unit lower mounting bolts. Place a block below the wheel to prevent the swinging arm dropping and the wheel spindle fouling the exhaust system.
2   Remove the brake caliper support bolt, which passes through the fork member on the left-hand side. Unscrew the spindle castellated nut after removing the split pin, and then draw the spindle out towards the left-hand side. Hold the caliper unit with one hand while removing the spindle.
3   Remove the spindle distance piece and lift the caliper unit up over the disc and away from the machine. Take care not to twist the hydraulic hose. Pull the rear-wheel over to the left until it disengages from the final drive gear. The wheel can now be lifted from the machine.
4   Replace the wheel by reversing the removal procedure. Always use a new split pin to secure the wheel spindle nut. The correct torque setting for the spindle nut is 58 - 72 lb. ft (8 - 10 kg . cm) or very tight. Grease the final drive spline with a lithium grease containing molybdenum disulphide additives.

7 Rear wheel bearings: examination and replacement

1   Remove the brake disc, which is retained by six nuts that screw on to studs in the hub. Lift out the final drive splined flange, the pins of which fit into cush drive rubber bushes. If the flange has not been removed for some time, corrosion between the pins and bush sleeves will cause difficulty in removal. A two-legged sprocket puller can be used to aid removal.
2   The bearing on the drive side can be drifted out from the brake side, using a cleft similar to that used for front wheel bearing removal, and the distance piece removed. The brake side bearing is retained by a screwed retainer ring, which should be removed using a peg spanner if possible. Drift out the bearing, together with the dust seal.
3   Check that bearings and seals and pack with grease as described for front wheel bearings in Section 4 of this Chapter. The remarks apply equally well. Replace the bearings by reversing the dimantling procedure.

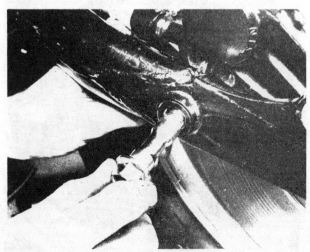

6 2a Unscrew brake caliper support bolt and ...

6.2b ... lift rear caliper out of position

6.3a Note wheel spacer between caliper and hub

6.3b Lift wheel across to left and remove

7.1a Rear wheel brake disc is retained on studs

7.1b Final drive/cush drive flange may need extracting

7.2 Screwed bearing retainer plate on disc side

7.3 Seal must face outwards when bearing is replaced

8 Front brake: replacing the pads and overhauling the caliper unit

1   Check the wear on the front brake pads, examine the pads through the small window in the main caliper units. If the red mark on the periphery of any pad has reached the brake.disc both pads in that set must be renewed. The rate of wear of the two sets are similar so it is probable that they will require renewal at the same time.

2   If the pads require renewal, remove each brake set individually, using the same procedure as follows. Unscrew the two large socket screws which clamp the caliper unit together. Pull the outer and inner sections of the caliper unit from position. The outer section is still attached to the hydraulic hose. Lift the old pads out, together with the shim.

3   Install the new pads and also the shim which fits against the outer face of the outer pad. The shim must be fitted so that the arrow is in the forward most position, pointing in an upward direction.

4   Refit the caliper halves and replace the socket screws. It may be necessary to push the caliper cylinder piston inwards to give the necessary clearance. If required, the bleed screw on the caliper can be slackened at the same time as the piston is pushed inwards. This

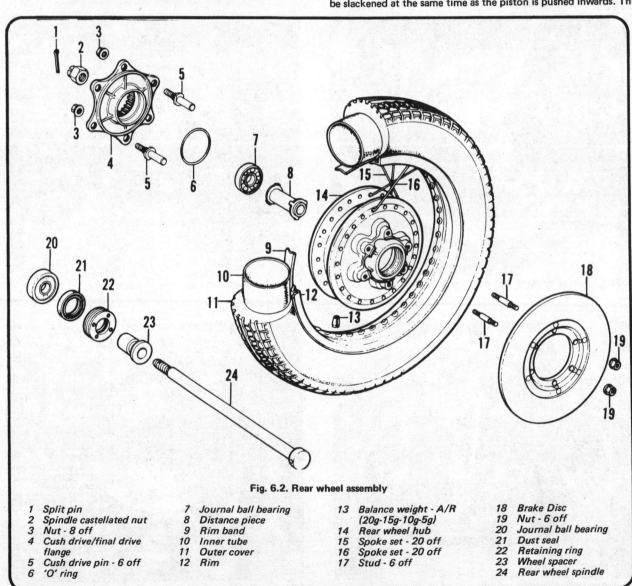

Fig. 6.2. Rear wheel assembly

1	Split pin	7	Journal ball bearing	13	Balance weight - A/R (20g-15g-10g-5g)	18	Brake Disc
2	Spindle castellated nut	8	Distance piece			19	Nut - 6 off
3	Nut - 8 off	9	Rim band	14	Rear wheel hub	20	Journal ball bearing
4	Cush drive/final drive flange	10	Inner tube	15	Spoke set - 20 off	21	Dust seal
5	Cush drive pin - 6 off	11	Outer cover	16	Spoke set - 20 off	22	Retaining ring
6	'O' ring	12	Rim	17	Stud - 6 off	23	Wheel spacer
						24	Rear wheel spindle

will allow a small amount of fluid to seep out and the piston to move easily. Place a rag around the bleed screw to prevent the fluid leaking onto the caliper unit.

5   Operate the brake lever after pad replacement, to check free movement of the pads and to allow the pads to self adjust.

6   When the caliper unit has been removed for pad replacement, it may be examined and overhauled as follows: Separate the caliper unit by removing the two inside socket screws. Note the position of the pad spring and lift it from the outer half of the caliper unit. Disconnect the hydraulic hose by unscrewing the banjo bolt, allowing the fluid to drain into a clean container.

7   Remove the boot clip and rubber boot from around the caliper piston. The piston can then be pulled out of the body. If necessary, apply air pressure to the fluid inlet to push the piston out.

8   Clean the brake components thoroughly and then check the condition of the cylinder and piston. If either component is scored or worn, both parts should be renewed. Pitting should also be looked for as this will reduce the sealing efficiency. Measure the outside diameter of the piston and the internal diameter of the cylinder. If either component is worn beyond the service limit, the parts must be renewed.

Piston outside diameter
service limit   1.5001 in (38.105 mm)

Cylinder internal diameter
service limit   1.5057 in (38.245 mm)

9   Hydraulic brake parts must be reassembled in absolutely clean conditions. Replace a new piston seal in the groove in the cylinder and grease the piston and seal with a high temperature silicon grease. If this is not available, brake fluid can be used. The piston must be refitted with the boot lip facing outwards. Replace the rubber boot and the retaining clip.

10 Check that the caliper pins move freely in the caliper anchor lubricating with silicon grease, is necessary. Check the rubber gaiters on the pins, which should be renewed if they have become perished.

11 Replace the caliper unit on the machine as for pad removal and replacement and bleed the brakes as described in Section 10. Check the brake operation carefully before the machine is used on the road.

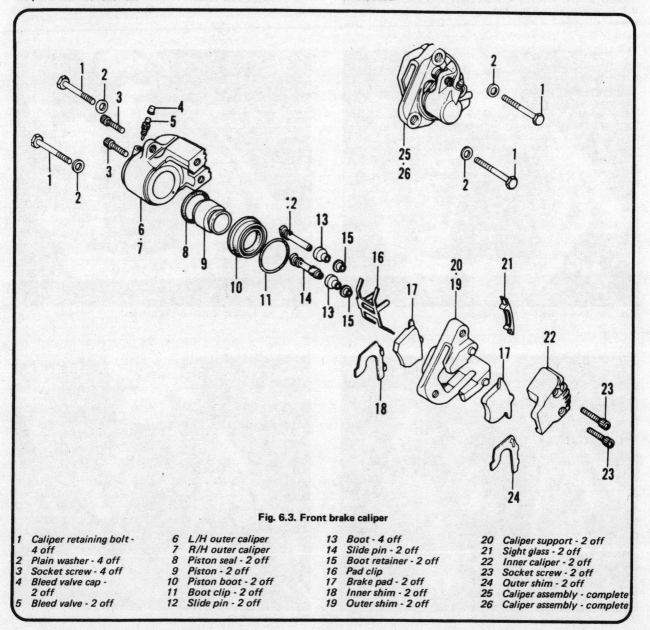

Fig. 6.3. Front brake caliper

1 Caliper retaining bolt - 4 off	6 L/H outer caliper	13 Boot - 4 off	20 Caliper support - 2 off
2 Plain washer - 4 off	7 R/H outer caliper	14 Slide pin - 2 off	21 Sight glass - 2 off
3 Socket screw - 4 off	8 Piston seal - 2 off	15 Boot retainer - 2 off	22 Inner caliper - 2 off
4 Bleed valve cap - 2 off	9 Piston - 2 off	16 Pad clip	23 Socket screw - 2 off
5 Bleed valve - 2 off	10 Piston boot - 2 off	17 Brake pad - 2 off	24 Outer shim - 2 off
	11 Boot clip - 2 off	18 Inner shim - 2 off	25 Caliper assembly - complete
	12 Slide pin - 2 off	19 Outer shim - 2 off	26 Caliper assembly - complete

8.2a Inner and outer caliper sections are retained by two allen screws

8.2b Remove inner casing (with wheel removed) to gain access ...

8 2c . . to inner brake pad followed by ...

8.2d ... the moving (outer) brake pad and ...

8.2e ... the caliper piston assembly

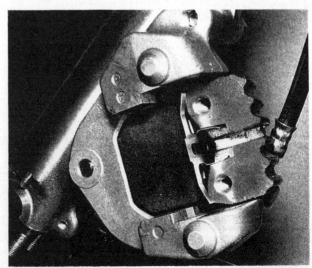

8.3a Note pad locating spring

8.3b Arrow on shim must be replaced as shown

9 Front master cylinder and hoses: examination and renovation

1 The master cylinder is unlikely to give trouble unless the machine has been stored for a long period or until a considerable mileage has been covered. The usual signs of trouble are leakage of fluid, causing a gradual fall in the fluid level, and bad braking performance. If dismantling should prove necessary, all work must be undertaken in clinically clean conditions.
2 To gain access to the master cylinder, commence the dismantling operation by attaching a bleed tube to one of the caliper bleed valves. Open the bleed valve one complete turn, then operate the front brake lever until all the hydraulic fluid is pumped out of the reservoir. There may be quicker methods of fluid draining but this is the least messey.
3 Remove the banjo bolt and hose from the end of the master cylinder unit and remove the lever pivot bolt and lever. Do not lose the rubber tube which is retained by the small plate on the pivot bolt. Cover the banjo end of the hydraulic hose to prevent the ingress of foreign matter, and tie the hose to the handlebars.
4 Unscrew the handlebar mirror and then detach the complete master cylinder assembly from the handlebars by removing the clamp. During removal, take great care not to spill any brake fluid onto painted surfaces or plastic.
5 The master cylinder can now be dismantled. Detach the rubber boot, after releasing the retaining clip. Note the backing washer on the boot. Removal of the internal circlip, which was obscured by the boot, will allow removal of the master cylinder components in the following order; Backing washer, piston with secondary cup, primary cup, spring and check valve.
6 Unscrew the two plate retaining screws inside the reservoir and lift the reservoir and plate away from the main cylinder body.
7 Take careful not of the way in which the seals are fitted since new ones must be replaced in the same order and position. Failure to observe this necessity will result in brake failure.
8 Clean the master cylinder with hydraulic fluid or alcohol. On no account use either abrasives or other solvents such as petrol Examine the condition of the piston, and more important, the cylinder bore, for damage or wear. It is not practicable to reclaim either the piston or the bore if damage has occured or if wear exceeds the following service limits.

Cylinder bore
Service limit (17.515 mm)

Piston outside diameter
Service limit 0.6850 in (17.400 mm)

9 Soak the new seals in hydraulic fluid for about 15 minutes prior to fitting, then reassemble the parts in exactly the same order, using the reversal of the dismantling procedure. Lubricate with hydraulic fluid and make sure the feathered edges of the seals are not damaged.
10 Refit the assembled master cylinder to the handlebars and reconnect the handlebar lever and hose. Refill the reservoir with hydraulic fluid and bleed the entire system by following the procedure described in Section 5 of this Chapter.
11 Check that the brake is working correctly before taking the machine on the road, to restore pressure and align the pads correctly. Use the brake gently for the first 50 miles or so to let the new components bed down correctly.
12 It should be emphasised that repairs to the master cylinder are best entrusted to a Honda dealer, or alternatively, that the defective parts should be replaced by a new unit. Dismantling and reassembly requires a certain amount of skill and it is imperative that the entire operation is carried out under surgically clean conditions.
13 Check the condition of all hydraulic hoses as part of routine safety checks made at regular intervals. If there is ANY doubt as to the condition of a hose due to abrasion, damage, splitting or bulging, it must be renewed immediately. If a hose is in a condition where failure is imminent, the failure will probably take place when the brakes are most needed, i.e. when the greatest pressure is applied.

10 Bleeding the brakes: front and rear systems

1 If for any reason, air has entered the hydraulic system, bleeding must take place to restore the correct functioning of the brake. Air is highly compressible and consequently if it becomes mixed with the hydraulic fluid in the system, it will prevent power being transmitted from the brake lever at the handlebars to the friction pad operating piston. Bleed the brake to remove the air as follows:
2 Attach a tube to the bleed valve at the top of the caliper unit, after removing the dust cap. It is preferable to use a transparent plastic tube, so that the presence of air bubbles can be seen more readily.
3 The far end of the tube should rest in a small bottle so that it is submerged in hydraulic fluid. This is essential to prevent air from passing back into the system. In consequence, the end of the tube must remain submerged at all times. Check that the reservoir on the handlebars is full of fluid and replace the cap to keep the fluid clean.
4 If spongy brake action necessitates the bleeding operation, squeeze and release the brake lever several times in rapid succession, to allow the pressure in the system to build up. Then open the bleed valve by unscrewing it slowly while still applying pressure to the brake lever. As soon as the pressure is felt to drop the bleed valve is sufficiently loosened. Squeeze the lever until it meets the handlebar and then tighten the bleed valve. If parts of the system have been replaced, the bleed valve can be left open from the beginning and the brake lever worked until fluid issues from the bleed tube. Note that it may be necessary to top up the reservoir during this operation; if it empties, air will enter the system and the whole operation will have to repeated.
5 Repeat operation 4 until bubbles disappear from the bleed tube. Close the bleed valve fully and replace the dust cap.
6 Check the level in the reservoir and top up if necessary. Never use the fluid which has drained into the bottle at the end of the bleed tube because this containins air bubbles which will be re-introduced into the system. It must stand for at least 24 hours before it can be reused.
7 Refit the diaphragm and diaphragm plate and tighten the reservoir cap securely.
8 Do not spill hydraulic fluid on the cycle parts. It is a very effective paint stripper! This also applies in the case of the speedometer and tachometer 'glasses', which it will attack.

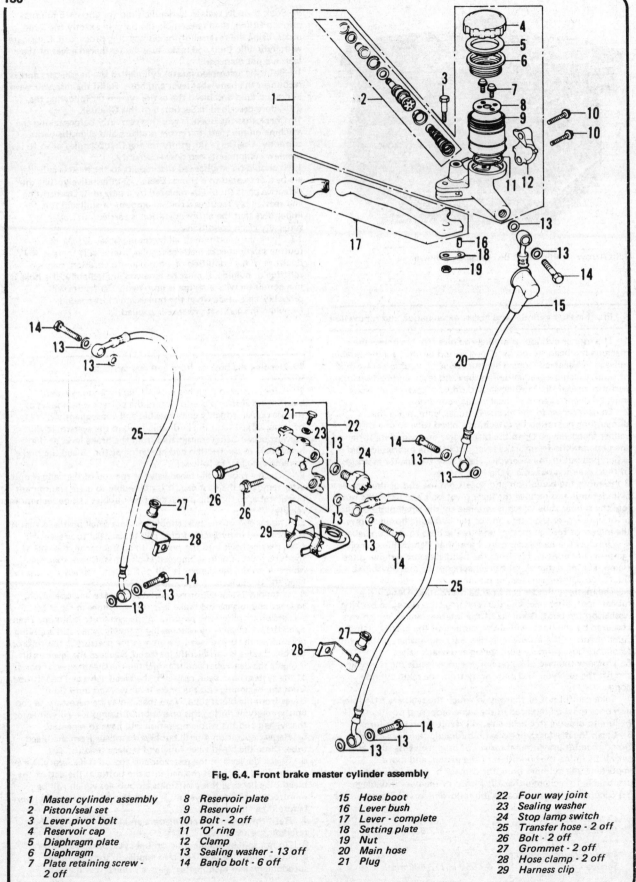

Fig. 6.4. Front brake master cylinder assembly

1 Master cylinder assembly	8 Reservoir plate	15 Hose boot	22 Four way joint
2 Piston/seal set	9 Reservoir	16 Lever bush	23 Sealing washer
3 Lever pivot bolt	10 Bolt - 2 off	17 Lever - complete	24 Stop lamp switch
4 Reservoir cap	11 'O' ring	18 Setting plate	25 Transfer hose - 2 off
5 Diaphragm plate	12 Clamp	19 Nut	26 Bolt - 2 off
6 Diaphragm	13 Sealing washer - 13 off	20 Main hose	27 Grommet - 2 off
7 Plate retaining screw - 2 off	14 Banjo bolt - 6 off	21 Plug	28 Hose clamp - 2 off
			29 Harness clip

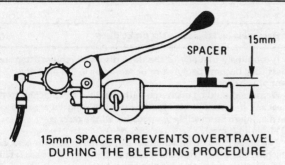

15mm SPACER PREVENTS OVERTRAVEL
DURING THE BLEEDING PROCEDURE

Fig. 6.5. Preventing brake lever overtravel during brake bleeding

11 Rear Brake:
replacing the pads and overhauling the caliper unit

1 Check rear brake pad wear after removing the plastic caliper
cover from position. If the red tongues on the pads have closed
together sufficiently that they are within the area marked red on
the caliper, they must be renewed.

2 Remove the caliper access cover, after unscrewing the single
screw. Depress the flat spring which locates both pad pins and
remove the upper pin. Tilt the spring back and pull out the lower
pin. The brake pad assembly can now be lifted from position.

3 Install a new set of pads together with the two shims, the
arrows on which must face downwards. Replace the pad pins,
ensuring that the pin locating spring locates correctly with the
'waisted' sections of the pins. Refit the cover and apply the brake
to adjust automatically. Note that the pads should always be
renewed as a pair.

4 Remove the caliper unit for overhauling in the same manner
as used for rear wheel removal (Section 6). Detach the caliper
from the hydraulic hose by unscrewing the banjo bolt. Either
allow the fluid from the hose to drain into a suitable receptacle
or tie the hose up at a level above that of the reservoir.

5 Remove the access cover and then separate the caliper unit
into two halves by removing the two socket screws. Remove the
pads and pins as a complete unit. Continue dismantling as
described for the front caliper units and refit parts and
reassemble in a similar manner. The service limits for the
various components are as follows:

Cylinder bore
Service limit   1.5057 in (38.245 mm)

Piston internal diameter
Service limit   1 5002 in (38.105 mm)

11.3b ... together on the locating spring

11.3c Insert them in the caliper unit and ...

11.3a Assemble the rear brake pads ...

11.3d ... refit the locating pins across the caliper unit

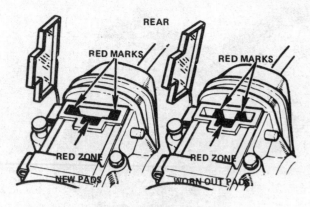

REAR

RED MARKS   RED MARKS

RED ZONE   RED ZONE

NEW PADS   WORN OUT PADS

Fig. 6.6. Rear brake pad replacement

12 Rear master cylinder: examination and renovation

1   Removal of the rear brake master cylinder can be accomplished as follows: Detach the right-hand side cover and disconnect the brake hose from the rear of the master cylinder by undoing the gland nut. Allow the fluid to drain into a suitable container while operating the brake pedal. Remove the split pin from the operating rod clevis pin and push the pin out of the fork.
2   The complete assembly can be detached by removing the two mounting bolts. Invert the master cylinder and unscrew the end fork, after slackening the locknut. Pull off the rubber boot. Extract the internal circlip and pull out the pushrod, followed by the remaining components in the following order. Piston and secondary cap, primary cap, spring, check valve.
3   Remove the two screws from the inside of the reservoir and pull the plate and reservoir off the body.
4   Check, renew and replace the components as described for the front brake master cylinder. The service limits are as follows:

Cylinder bore
Service limit   1.5533 in (14.055 mm)

Piston outside diameter
Service limit   1.5057 in (13.940 mm)

12.1 Rear master cylinder retained by two bolts

13 Removing and replacing the brake discs

1   It is unlikely that any of the three discs will require attention unless they become badly worn or scored or in the unlikely event of warpage.
2   Warpage should be measured with the discs still atached to the wheel and the wheel in situ on the machine using a dial gauge. The maximum permissible warpage for all the discs is 0.0118 in (0.3 mm).
3   The brake discs on either wheel can be removed with ease, after detaching the wheel from the machine. The front discs are retained by long bolts passing through the hub which, retain also the speedometer drive dog retainer plate. The rear disc is retained on studs by self locking nuts.
4   The brake discs will wear eventually to thickness which no longer allow sufficient support, and will probably begin to warp. The correct wear service limits which can be measured with a micrometer are as follows:

Front disc minimum thickness   0.1969 in (5.0 mm)
Rear disc minimum thickness   0.2362 in (6.0 mm)

13.4 Disc wear limit is marked on disc

14 Cush drive assembly: examination

1   A cush drive assembly is incorporated in the rear wheel to absorb any shocks transmitted from the final drive gear to the wheel. The system comprises six 'flexible' bushes inserted in the rear hub into which fit the pins of the final drive flange, which is connected to the drive gear by a splined boss.
2   After considerable service the bushes will wear and become compacted, giving rise to excessive backlash between the drive shaft and the wheel. Since the bushes are not available as separate replacement parts, this will mean that the complete rear hub must be renewed. The only alternative to this somewhat expensive solution is to find a good Honda dealer who is prepared to compare the bushes with those in his stock; many Honda machines employ a similar cush-drive arrangement, and one or two may use bushes of the same size and strength that are available separately.
3   If bushes cannot be found, a new hub must be purchased and fitted by a wheel-building expert. However, if replacement bushes are found the drive flange can be pulled from the hub after the wheel has been removed. If the flange pins are stuck by corrosion in the bush centres, a two or three legged sprocket puller can be used to extract the flange, as shown in photograph 7.1b. The bushes themselves are notoriously difficult to remove; it is recommended that the wheel be taken to a good Honda dealer for removal to be accomplished using the correct internally expanding bearing puller. New bushes can be drifted into place after lubricating both inner and outer sleeves with a smear of graphite grease.

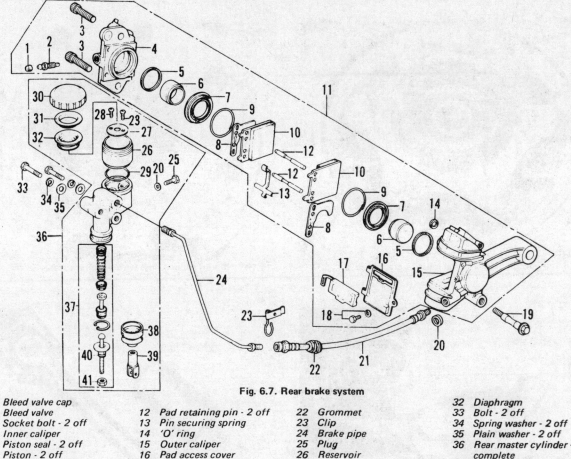

Fig. 6.7. Rear brake system

1	Bleed valve cap	12	Pad retaining pin - 2 off
2	Bleed valve	13	Pin securing spring
3	Socket bolt - 2 off	14	'O' ring
4	Inner caliper	15	Outer caliper
5	Piston seal - 2 off	16	Pad access cover
6	Piston - 2 off	17	Inspection window
7	Piston boot - 2 off	18	Bolt
8	Shim - 2 off	19	Caliper support bolt
9	Boot clip - 2 off	20	Sealing washer - 2 off
10	Pad set - 2 off	21	Hose
11	Caliper assembly complete		

22	Grommet	32	Diaphragm
23	Clip	33	Bolt - 2 off
24	Brake pipe	34	Spring washer - 2 off
25	Plug	35	Plain washer - 2 off
26	Reservoir	36	Rear master cylinder - complete
27	Reservoir plate	37	Piston/seal set
28	Reservoir plate screw - 2 off	38	Boot
29	'O' ring	39	Rod fork
30	Reservoir cap	40	Piston rod
31	Diaphragm plate	41	Locknut

15 Final drive system: examination and renovation

1   The final drive housing and drive shaft can be removed from the machine by following the procedure given for swinging arm removal in Chapter 5, Section 10. The only time when the swinging arm does not require removal to extract the final drive shaft is when the engine unit is out of the frame

2   It is strongly advised that the final drive housing incorporating the final drive crown and pinion gears be returned to a Honda service agent if and when servicing or overhaul is required. Dismantling the gear unit requires the use of special tools which are not generally obtainable by the public. Additionally, reassembly of the unit requires working to very close tolerances and the preload adjustment of bearings and gears.

3   Check the drive shaft after removal for slop in the splined joints and for wear in the universal joint. Renew the matching components, where necessary

16 Tyres: removing and replacing

1   At some time or other the need will arise to remove and replace the tyres, either as the result of a puncture or because a renewal is required to offset wear. To the inexperienced, tyre changing represents a formidable task, yet if a few simple rules are observed and the technique learned, the whole operation is surprisingly simple.

2   To remove the tyre from either wheel, first detach the wheel from the machine by following the procedure in Sections 3 or 6, depending on whether the front or the rear wheel is involved. Deflate the tyre by removing the valve insert and when it is fully deflated, push the bead of the tyre away from the wheel rim on both sides so that the bead enters the centre well of the rim. Remove the locking cap and push the tyre valve into the tyre itself.

3   Insert a tyre lever close to the valve and lever the edge of the tyre over the outside of the wheel rim. Very little force should be necessary; if resistance if encountered it is probably due to the fact that the tyre beads have not entered the well of the wheel rim all the way round the tyre.

4   Once the tyre has been edged over the wheel rim, it is easy to work around the wheel rim so that the tyre is completely free on one side. At this stage, the inner tube can be removed.

5   Working from the other side of the wheel, ease the other edge of the tyre over the outside of the wheel rim which is furthest away. Continue to work around the rim until the tyre is free completely from the rim.

6   If a puncture has necessitated the removal of the tyre, re-inflate the inner tube and immerse it in a bowl of water to trace the source of the leak. Mark its position and deflate the tube. Dry the tube and clean the area around the puncture with a petrol soaked rag. When the surface has dried, apply the rubber solution and allow this to dry before removing the backing from the patch and apply the patch to the surfaces.

7   It is best to use a patch of the self-vulcanising type which will form a very permanent repair. Note that it may be necessary to remove a protective covering from the top surface of the patch,

15.1a Splines on final drive flange must be greased as ...

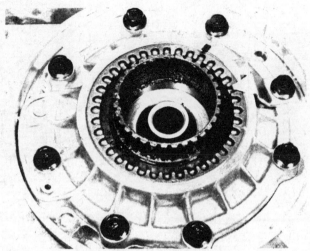

15.1b ... should splined boss with which it locates

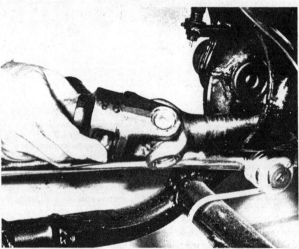

15.1c Final drive shaft removal (with engine removed)

15.3a Check U-joint for wear and also ...

15.3b ... mating splines on final drive pinion shaft

after it has sealed in position. Inner tubes made from synthetic rubber may require a special type of patch and adhesive if a satisfactory bond is to be achieved.

8   Before refitting the tyre, check the inside to make sure that the object which caused the puncture is not trapped. Check the outside of the tyre, particularly the tread area, to make sure nothing is trapped that may cause a further puncture.

9   If the inner tube has been patched on a number of past occasions, or if there is a tear or large hole, it is preferable to discard it and fit a new one. Sudden deflation may cause an accident, particularly if it occurs with the front wheel.

10 To replace the tyre, inflate the inner tube sufficienctly for it to assume a circular shape but only just. Then push it into the tyre so that it is enclosed completely. Lay the tyre on the wheel at an angle and insert the valve through the rim tape and the hole in the wheel rim. Attach the locking ring on the first few threads, sufficient to hold the valve captive in its correct location.

11 If the tyre has a balance mark (usually a dot), as on the tyres fitted as original equipment, this must be positioned alongside the valve. Similarly, any arrow indicating direction of rotation must face the right way.

12 Starting at the point furthest from the valve, push the tyre bead over the edge of the wheel rim until it is located in the central well. Continue to work around the tyre in this fashion until the whole of one side of the tyre is on the rim. It may be

Tyre changing sequence - tubed tyres

A Deflate tyre. After pushing tyre beads away from rim flanges push tyre bead into well of rim at point opposite valve. Insert tyre lever adjacent to valve and work bead over edge of rim.

B Use two levers to work bead over edge of rim. Note use of rim protectors

C Remove inner tube from tyre

D When first bead is clear, remove tyre as shown

E When fitting, partially inflate inner tube and insert in tyre

F Work first bead over rim and feed valve through hole in rim. Partially screw on retaining nut to hold valve in place.

G Check that inner tube is positioned correctly and work second bead over rim using tyre levers. Start at a point opposite valve.

H Work final area of bead over rim whilst pushing valve inwards to ensure that inner tube is not trapped

necessary to use a tyre lever during the final stages.

13 Make sure that there is no pull on the tyre valve and again commencing with the area furthest from the valve, ease the other bead of the tyre over the edge of the rim. Finish with the area close to the valve, pushing the valve up into the tyre until the locking cap touches the rim. This will ensure the inner tube is not trapped when the last section of the bead is edged over the rim with a tyre lever.

14 Check that the inner tube is not trapped at any point. Re-inflate the inner tube and check that the tyre is seating correctly around the wheel rim. There should be a thin rib moulded around the wall of the tyre on both sides which should be equidistant from the wheel rim at all points. If the tyre is unevenly located on the rim, try bouncing the wheel when the tyre is at the recommended pressure. It is probable that one of the beads has not pulled clear of the centre well.

15 Always run the tyres at the recommended pressures and never under or over-inflate. The correct pressures for solo use are given in the Specification Section of this Chapter. If a pillion passenger is carried, increase the rear tyre pressure only as recommended.

16 Tyre replacement is aided by dusting the side walls, particularly in the vicinity of the beads, with a liberal coating of French chalk. Washing up liquid can also be used to good effect. but this has the disadvantage of causing the inner surfaces of the wheel rim to rust.

17 Never replace the inner tube and tyre without the rim tape in position. If this precaution is overlooked there is good chance of the ends of the spoke nipples chafing the inner tube and causing a crop of punctures.

18 Never fit a tyre which has a damaged tread or side walls. Apart from the legal aspects, there is a very great risk of blow-out, which can have serious consequences on any two-wheel vehicle.

19 Tyre valves rarely give trouble, but it is always advisable to check whether the valve itself is leading before removing the tyre. Do not forget to fit the dust cap, which forms an effective second seal. This is especially important on high performance machines where centrifugal force can cause the valve insert to retract and the tyre to deflate without warning.

17 Valve cores and caps

1 Valve cores seldom give trouble, but do not last indefinitely. Dirt under the seating will cause a puzzling 'slow puncture'. Check that they are not leaking by applying spittle to the end of the valve, and watching for air bubbles.

2 A valve cap is a safety device, and should always be fitted. Apart from keeping dirt out of the valve, it provides a second seal in case of valve failure, and may prevent an accident resulting from sudden deflation.

18 Front wheel balancing

1 The front wheel should be statically balanced, complete with tyre. An out of balance wheel can produce dangerous wobbling at high speed.

2 Some tyres have balance mark on the sidewall. This must be positioned adjacent to the valve. Even so, the wheel still requires balancing.

3 With the front wheel clear of the ground, spin the wheel several times. Each time, it will probably come to rest in the same position. Balance weights should be attached diametrically opposite the heavy spot, until the wheel will not come to rest in any set position, when spun.

4 Balance weights, which clip round the spokes, are available in 5, 10, or 20 grammes weight. If they are not available, wire solder wrapped round the spokes and secured with insulating tape will make a substitute.

5 It is possible to have a wheel dynamically balanced at some dealers. This requires its removal.

6 There is no need to balance the rear wheel under normal road conditions, although any tyre balance mark should be aligned with the valve.

19 Fault diagnosis: wheels, brakes and tyres

Symptom	Cause	Remedy
Handlebars oscillate at low speeds	Buckled front wheel Incorrectly fitted front tyre	Remove wheel for specialist attention Check whether line around bead is equidistant from rim
Forks 'hammer' at high speeds	Front wheel out of balance	Add weights until wheel will stop in any position
Brakes grab, locking wheel	Ends of brake shoes not chamfered (rear)	Remove brake shoes and chamfer ends
Brakes feel spongy	Badly distorted disc (front). Weak pull off springs (rear) Air has entered hydraulic system (front)	Check and renew disc if necessary. Renew springs after inspection Bleed system
Tyres wear more rapidly in middle of tread	Over-inflation	Check pressures and run at recommended settings
Tyres wear rapidly at outer edge of tread	Under-inflation	Ditto.

19 Fault diagnosis: wheels, brakes and tyres

Symptom	Cause	Remedy
Handlebars oscillate at low speeds	Buckled front wheel Incorrectly fitted front tyre	Remove wheel for specialist attention Check whether line around bead is equidistant from rim
Forks 'hammer' at high speeds	Front wheel out of balance	Add weights until wheel will stop in any position
Brakes grab, locking wheel	Ends of brake shoes not chamfered (rear)	Remove brake shoes and chamfer ends
Brakes feel spongy	Badly distorted disc (front). Weak pull off springs (rear) Air has entered hydraulic system (front)	Check and renew disc if necessary. Renew springs after inspection Bleed system
Tyres wear more rapidly in middle of tread	Over-inflation	Check pressures and run at recommended settings
Tyres wear rapidly at outer edge of tread	Under-inflation	Ditto.

Chapter 7 Electrical system

Contents

Specifications

Earth ... ...   Negative

Battery
Type ... ...   Y50 - N18L - A2
Make ... ...   YUASA
Voltage 12v
Capacity ... ...   20 AH
Alternator output ... ...   300 watt @5,000 rpm

Bulbs*
Headlight ... ...   Tungsten incandescent 40/50W
Tail stop light 8/27W
Position/front indicator 8/23W
Rear indicator 23W
Instruments 3.4W
Indicator tell-tale 3.4W

Fuses
Main (battery) 30A
Headlight 10A
Tail/instrument lights 5A
Parking (tail light) 5A
Oil/temp/neutral/fuel 5A
Horn/stop lamp/indicators 15A

All bulbs are rated at 12 volts, however, some variation may occur in the wattage rating, depending on the statutory requirements of the country or state to which the machine was first delivered.

Starter motor
Brush length:

Standard ...	...	...	...	...	...	...	...	0.4724 - 0.5118 in (12 - 13 mm)
Service limit ...	...	...	...	...	...	...	...	0.2165 in (5.5 mm)

1 General description

1 The Honda Gold Wing is fitted with a 12v negative earth electrical system. Power is provided by a three-phase fixed-coil alternator, mounted on a cush-drive double gear shaft, which is driven from the crankshaft. The AC (alternating current) output from the alternator is converted to DC (direct current) by a silicon rectifier and is controlled by a solid state voltage regulator. In addition to the main lighting and indicating system a reserve system is incorporated on all machines supplied to the American market. This system functions as follows. If the filaments of either 'main' or 'dip' beam burn out while the headlamp is on, the system automatically switches power over to the undamaged filament, at the same time indicating the failure by illuminating a warning light in the console. If the main beam filament fails the headlamp is automatically switched onto 'dip' beam, at full power. If the 'dip' beam filament fails the main beam is switched on at 45% of the normal rated voltage. If the dipswitch is then moved to the 'main' beam position the main beam filament is illuminated at full power. The stop lamp is switched to 45% power if the tail light fails. The 'tail light' indicator on the console simultaneously being switched from 'on' to 1/3 power. If the brakes are then applied the stop lamp switches to full power as does the 'tail light' indicator. Additionally the 'tail ight' indcator ceases to function as a warning if the stop lamp fails. The system utilises an oscillator which senses bulb failure, and a transistorised switching unit which provides selective control of light bulbs and warning lamps. Two resistance units are incorporated to reduce the power as described above, to 40% - 50% of full power.

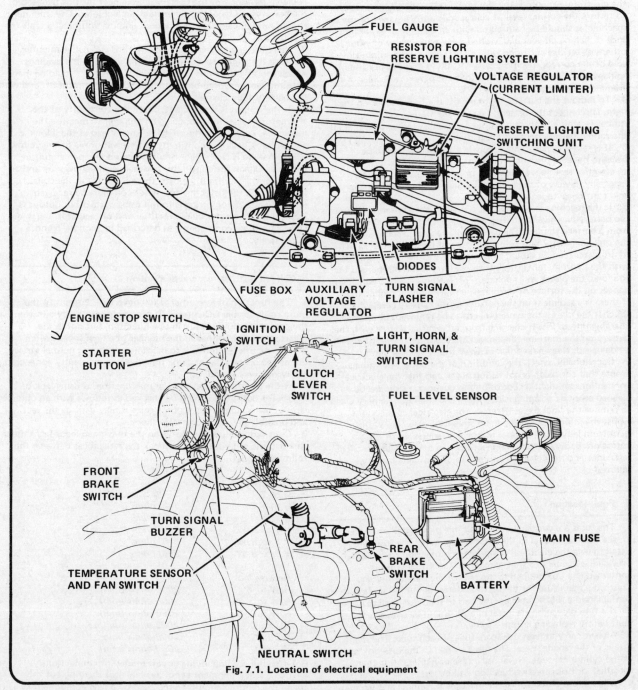

Fig. 7.1. Location of electrical equipment

2 Battery: examination, topping-up and maintenance

1 The battery is housed behind the right-hand side cover, and is fixed with a rubber strap.

2 The transparent plastic case of the battery permits the upper and lower levels of the electrolyte to be observed without disturbing the battery by removing the side cover. Maintenance is normally limited to keeping the electrolyte level between the prescribed upper and lower limits and making sure that the vent tube is not blocked. The lead plates and their separators are also visible through the transparent case, a further guide to the general condition of the battery. If electrolyte level drops rapidly, suspect over-charging and check the system.

3 Unless acid is split, as may occur if the machine falls over, the electrolyte should always be topped up with distilled water to restore the correct level. If acid is spilt onto any part of the machine, it should be neutralised with an alkali such as washing soda or baking powder and washed away with plenty of water, otherwise serious corrosion will occur. Top up with sulphuric acid of the correct specific gravity (1.260 to 1.280) only when spillage has occurred. Check that the vent pipe is well clear of the frame or any of the other cycle parts.

4 To remove the battery, first take off the frame cover on the right. Disconnect the battery, take off the frame rubber retaining strap, and lift out the battery. Make sure that all the rubber battery pads are in position.

5 It is seldom practicable to repair a cracked battery case because the acid present in the joint will prevent the formation of an effective seal. It is always best to renew a cracked battery, especially in view of the corrosion which will be caused if the acid continues to leak.

6 If the machine is not used for a period, it is advisable to remove the battery and give it a 'refresher' charge every six weeks or so from a battery charger. The battery will require recharging when the specific gravity falls below 1.2 per cell (at 20°C - 68°F). The hydrometer reading should be taken at the top of the meniscus with the hydrometer vertical. If the battery is left discharged for too long, the plates will sulphate. This is a grey deposit which will appear on the surface of the plates, and will inhibit recharging. If there is a sediment on the bottom of the battery case, which touches the plates, the battery needs to be renewed. Limit the charging rate to 2A. If charging from an external source with the battery on the machine, disconnect the leads, or the rectifier will be damaged. Keep naked flames away from the filler vents.

7 Occasionally, check the condition of the battery terminals to ensure that corrosion is not taking place, and that the electrical connections are tight. If corrosion has occurred, it should be cleaned away by scraping with a knife and then using emery cloth to remove the final traces. Remake the electrical connections whilst the joint is still clean, them smear the assembly with petroleum jelly (NOT grease) to prevent recurrence of the corrosion. Badly corroded connections can have a high electrical resistance and may give the impression of complete battery failure.

3 Fuse: location

1 The fuses are contained in a bank in a plastic box, mounted on the electrical panel in the dummy fuel tank. The fuses are fitted to give the electrical components protection from sudden overload as occurs in a short circuit. An additional fuse is contained in a box clipped to the rear of the battery holder. This is the main fuse for the battery/starter circuit. A spare fuse is contained in a rubber pocket attached to the battery strap.

2 If a fuse blows the electrical circuit should be checked for a fault before replacing it with another.

3 Always carry at least one spare fuse of each type. The main fuse is of the 'spade' type. The fuses carried in the fuse box are of the cylindrical type. Never use a fuse of higher rating than specified or its protective function will be lost.

4 When a fuse blows whilst the machine is running and no spare fuse is available a 'get you home' remedy is to remove the blown fuse and wrap it in silver paper, this will restore the electrical continuity by bridging the broken wire within the fuse. This expedient should not be used if there is evidence of a short circuit or major electrical fault, otherwise more serious damage will be caused. Replace the 'doctored' fuse at the earliest possible opportunity to restore full circuit protection.

4 Alternator : checking output and continuity

1 If the output of the charging system is suspect, it may be checked as follows, using an ammeter and a voltmeter, connected into the circuit. Connect the voltmeter across the positive and negative terminals of the battery. Disconnect the positive lead of the battery and reconnect it through an ammeter. The ammeter should have a scale of 0 - 5 amperes, and the voltmeter a scale of 0 - 20 volts.

2 Start the engine and allow it to run for about five minutes. Increase the engine speed to 5,000 rpm and take the readings which should be 3 amps and 14.5 volts. This test should be made with the dipper switch on 'main' beam, the fan motor off and with the battery fully charged.

3 If the output is insufficient, check the continuity of the alternator coils as follows, using an ohmeter to measure the resistance. Disconnect the alternator main lead at the 'block' c connector. Check the continuity between all three of the yellow wires. If there is lack of continuity, there is an open circuit in the coils. Check the continuity between each yellow wire and an earth point. If continuity exists, there is a short in the coils. In either case the defective alternator coil must be renewed.

4 If the alternator wiring is found to be correct but output is incorrect, suspect the silicon rectifier and/or regulator. Carry out checks on these components as described in Sections 5 and 6 respectively.

5 Silicon rectifier: general description

1 The function of a rectifier is to convert AC current to the more easily usable DC current. It accomplishes this by allowing the current to flow readily in one direction, but not in the reverse direction, rather in the manner of a high speed switch.

2 The rectifier is unlikely to malfunction during normal service, unless the battery has at some time been inadvertently connected the wrong way round.

If the rectifier is suspected of malfunction, causing lack of alternator output, it should be checked as follows with an ohmeter. Do not use a high voltage source since it may damage the rectifier and possibly give the operator a shock.

3 Continuity must exist between the two green leads from the unit, and also the red/white leads. Check the resistances between the following leads.

Forward bias	
G to Y 1	5 - 40 ohms
G to Y 2	5-40 ohms
G to Y 3	5-40 ohms
Y 1 to R/W	5-40 ohms
Y 2 to R/W	5-40 ohms
Y 3 to R/W	5-40 ohms

Reverse bias	
R/W to Y 1	2000 ohms (min)
R/W to Y 2	2000 ohms (min)
R/W to Y 3	2000 ohms (min)
Y 1 to G	2000 ohms (min)
Y 2 to G	2000 ohms (min)
Y 3 to G	2000 ohms (min)

Refer to the accompanying circuit model for connections.

4 If the test equipment is not available for carrying out this check the rectifier should be replaced by a new component as a

method of checking by direct replacement. If the rectifier is found to be malfunctioning it must be renewed as a repair is impracticable.

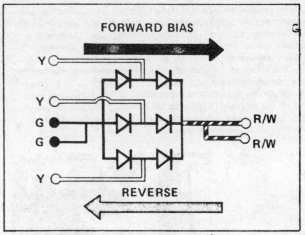

Fig. 7.2. Checking the silicon rectifiers

3.1 Main fuse is of the 'spade' type

4.1 Alternator stator coils are retained by three screws

5.1 Silicon rectifier is attached to battery carrier

6 Voltage regulator : checking routine

1   A non-adjustable solid state voltage regulator is used to maintain a stable output voltage from the alternator. The unit is fitted to the electrical panel within the dummy fuel tank. If the unit malfunctions, a replacement will have to be obtained, as no repair is possible.

2   Test the regulator as follows. Connect a voltmeter of 0 - 20 volt range accross the positive and negative terminals of the battery. Connect an ammeter between the green regulator lead and an earth point. Start the engine. The regulator must divert the current through the ammeter to earth when the battery voltage reaches 14 - 15 volts.

7 Starter motor: removal, examination and replacement

1   A push button switch is located on the right-hand section of the handlebars. When depressed, it operates a solenoid which in turn causes the starter motor to operate. This drives a free running clutch via a chain, which in turn rotates an idler shaft upon which the alternator is mounted, and finally the crankshaft.

2   To remove the starter motor, first detach the negative terminal at the battery and the main lead on the starter motor body. Remove the left-hand exhaust pipe and the gear change lever. The starter motor is retained by two bolts passing into the crankcase. After removal of the bolts the motor can be eased backwards and away from the machine.

3   To remove the brushes first unscrew the two long external screws on the outside of the starter motor body. The end cover can then be removed, exposing the commutator and brushes .

4   Lift up the spring clips retaining the brushes and remove the brushes from their holders. Measure their length with a pair of vernier calipers and check the service wear limit in the Specifications Section at the front of this Chapter.

5   Before replacing the brushes, make sure the commutator is clean. Clean with a strip of fine emery cloth, followed by metal polish, then wipe with a rag soaked in petrol to ensure a grease free surface.

6   Reassemble in the reverse order, making sure that the brushes can move freely in their holders.

8 Starter motor switch: function, removal and examination

1   When the handlebar starter button is depressed the solenoid switch engages which in turn makes the circuit between the battery and starter. A solenoid switch is necesary to withstand the heavy current to produce the starting torque (about 120 amps) which would otherwise overload the handlebar button if applied direct.

2　To remove the solenoid, detach the left-hand side cover and disconnect the battery leads. Disconnect the two leads from the starter solenoid which run to the handlebar lever. They are connected by 'snap' connectors. Remove the two main leads which are retained on the switch by nuts and washers.

3　The solenoid is a push fit in a rubber mounting, attached to the rear of the battery corner.

4　When the starter button is pressed, a click should be heard inside the solenoid. This indicates that the points inside are closing. No repair is possible so if the switch malfunctions a replacement must be obtained.

8.2 Starter solenoid is protected by rubber cover

9　Ignition switch: removal and replacement

1　If the ignition switch fails no repair is possible. A new component must be obtained.

2　To remove the switch it will first be necessary to detach the instrument heads, complete with warning panel and bracket. This can be accomplished after disconnecting the drive cables and removing the two instrument bracket mounting bolts.

3　Disconnect the wires to the ignition switch, noting the position of the coloured wires. On reassembly, refer to the accompanying ignition switch wiring diagram. After removal of the two bolts that retain the switch from below the upper fork yoke the complete switch can be removed.

4　Refitting of the switch should take place by reversing the removal procedure.

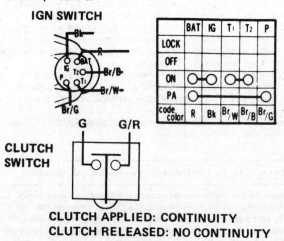

Fig. 7.3. Clutch switch - continuity test

10　Handlebar switches

1　To remove the handlebar switches, first disconnect the battery and also the leads to the switches, which are located either within the headlamp shell or in the electrical compartment of the dummy fuel tank.

2　Both switches can be detached from the handlebars after separating the halves which are clamped together from underneath by two cross-head screws each.

3　The throttle cables will have to be disconnected from the right-hand switch. This can be accomplished by slackening the adjusters off at both ends to give the required amount of play.

4　Disconnect the clutch cable from the left-hand handlebar switch assembly. Note also the small switch on the clutch lever which isolates the solenoid and so prevents starting of the engine when the machine is in gear and the clutch is not disengaged.

5　Remove the two clamps which retain the handlebars to the upper yoke. Each clamp is retained by two nuts.

6　The leads from the main switches run through the centre of the handlebars, to emerge from the slot in the central forward part of the handlebars. Before pulling the wires from the handlebars attach a thin length of wire to the wire ends. The thin wire will be drawn through and can be used to pull the main wires back through the bars, during replacement.

7　Repair of either switch assembly is seldom practicable. If the switches malfunction due to corrosion or dirt on the contacts they can be cleaned most easily with a proprietary electrical contact cleaner which can be sprayed onto the affected areas. It is possible to dismantle the switches, but reassembly will be a problem unless the operator is an accomplished watch maker! The contact components are minute.

11　Stop lamp switch: removal and adjustment

1　The stop lamp switch is a pull-type switch held in a bracket on the frame and operated by the brake pedal.

2　If the setting is not correct the nut on the body of the stop lamp should be adjusted in the appropriate direction (this will be necessary after the rear brake itself has been adjusted.

3　Raising the switch body by turning the adjusting nut clockwise will make the stop lamp operate earlier. Conversely, if the nut is turned anticlockwise, the switch is lowered and the stop lamp operates later.

4　As a guide, the stop lamp should operate after the brake pedal has been depressed by ¾ in. A front brake lamp switch is incorporated in the system, fitted into the junction piece on the lower fork yoke. Renewal of the switch requires bleeding of the front brakes, after the new switch has been fitted. When unscrewing the damaged switch, place a container beneath the junction to prevent spillage of hydraulic fluid.

12　Neutral indicator switch: removal and replacement

1　The neutral indicator switch is fitted in the right-hand wall of the crankcase, forward of the oil strainer access cover.

2　If it is not functioning correctly (after checking that the warning bulb is not at fault) the switch must be renewed. Drain the engine oil to lower the level in the crankcase, and remove the right-hand exhaust pipe so that access to the switch can be gained. Remove the bolt and claw plate that retains the switch and remove the front right-hand engine mounting nut. In order that the switch clears the engine mounting lug the frame must be prised outwards about 0.08 in (2.0 mm). This can be accomplished by placing a wooden lever between the engine and frame. Pull the switch out until it leaves the crankcase.

3　Fit a new switch by reversing the removal procedure and reconnect.

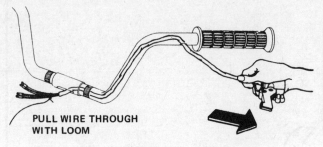

PULL WIRE THROUGH
WITH LOOM

Fig. 7.4. Tracking switch wiring through handlebars

13.2a Headlamp rim is secured by three screws

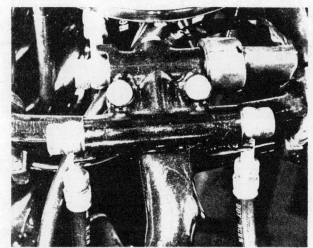

11.4 Front stop lamp switch fitted in hose junction

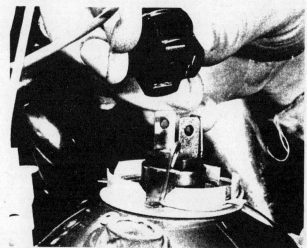

13.2b Headlamp bulb connected by plug and socket

13 Headlamp: replacement of bulbs

1   Depending on the model, the headlamp bulb is either held in
a separate bulb holder, which fits into the rear of the reflector, or
is an integral part of the reflector and therefore a sealed beam unit.
2   In both cases initial access is gained by removing the headlamp
rim, complete with glass and reflector, which is retained by three
screws through the headlamp shell.
3   Removal of the detachable bulb type is quite straightforward.
Pull off the retaining spring and prise the holder from position.
The headlamp bulb is integral with the holder and is non-detachable.
The pilot bulb (where fitted) is either held in the main bulb
holder, and is of the torpedo type and can be prised from position
or is retained in a separate bulb holder below the main holder.
In this case pull the holder from position and push and twist the
bayonet fixed bulb for removal.
4   If a sealed beam headlamp become defective it is necessary to
renew the complete glass/reflector unit as the bulb element is not
detachable. The unit is held by screws passing through tabs on the
headlamp rim. It will also be necessary to disconnect the horizontal
beam adjustment which comprises a screw and tensioner spring. No
pilot bulb is fitted to sealed beam units.

13.3a Bulb is retained in reflector by spring and ...

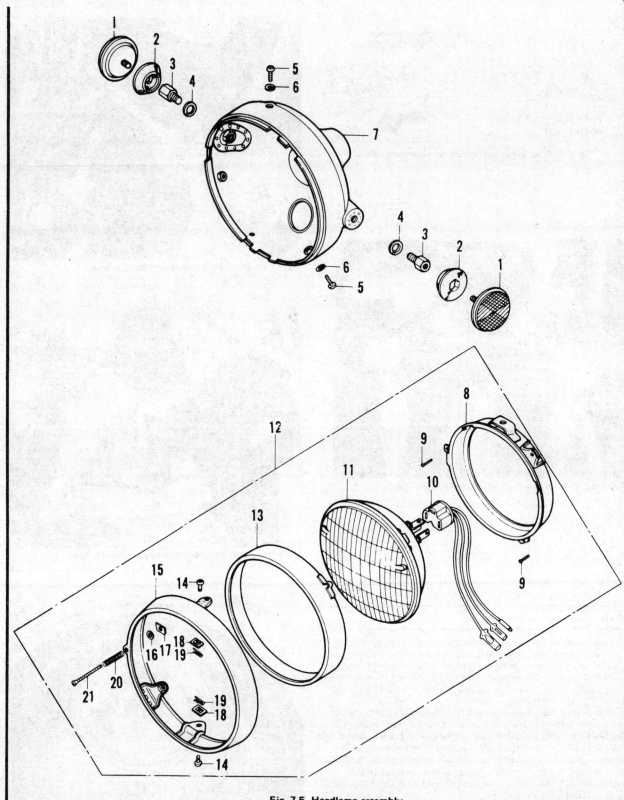

Fig. 7.5. Headlamp assembly

1 Reflector - 2 off	7 Headlamp shell	12 Headlamp reflector unit - complete	16 Washer
2 Reflector seat - 2 off	8 Lens holder		17 Adjuster nut
3 Support bolt - 2 off	9 Screw - 3 off	13 Inner reflector ring	18 Pivot clip - 2 off
4 Spring washer - 2 off	10 Socket	14 Pivot screw - 2 off	19 Spring clip - 2 off
5 Screw - 3 off	11 Sealed beam	15 Headlamp rim	20 Spring
6 Washer			21 Adjuster screw

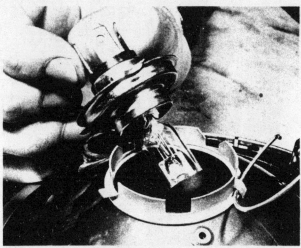

13.3b ... is a push fit in reflector centre

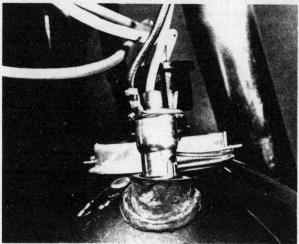

13.3c Pilot bulb holder is a push fit and ...

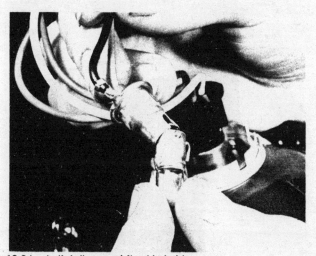

13.3d ... bulb is 'bayonet' fixed in holder.

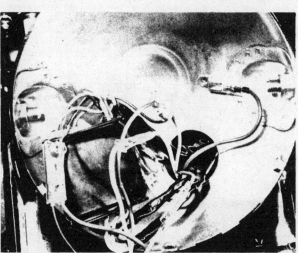

13.3e Main wiring is colour coded

14 Headlamp: adjusting beam height

1   The beam height of the headlamp is adjusted by slackening the mounting bolts and tilting if forward or backwards as required.
2   To set up the headlamp the machine should be placed on level ground at least 25 feet away from a wall in its normal position, ie. off its stand and with a rider (also a pillion passenger if one is normally seated). On main beam the height of the centre of the headlight from the ground should be the same. This will ensure that when the dip switch is operated the lamp will not give a dazzling effect.
3   Horizontal beam adjustment can only be made on sealed beam headlamp units. A screw passing through the headlamp rim on the right-hand side allows the complete glass/reflector unit to be moved about a vertical pivot. Screw the adjuster in clockwise to raise the beam and vice versa.
4   UK lighting regulations stipulate that the lighting system must be arranged so that the light will not dazzle a person standing at a distance greater than 25 feet from the lamp, whose eye level is not less than 3 ft 6 inches above that plane. It is easy to approximate this setting by placing the machine 25 feet away from a wall, on a level road and setting the beam height so that it is concentrated at the same height as the distance of the

centre of the headlamp from the ground. The rider must be seated normally during this operation and also the pillion passenger, if one is carried regularly.

15 Stop and tail lamp: replacing bulbs

1   The rear lamp has a twin filament bulb. The stop lamp operates when both front and rear brakes are applied. In the UK, a working stop light is a statutory requirement.
2   To renew the bulb, first remove the lens after unscrewing the cross-head screws. Check the rubber gasket underneath the lens. The bulb is bayonet fitting, with offset pins so that it can be replaced in one position only. Do not touch the glass envelope; use a cloth or tissue. Make sure that the contacts are clean.
3   When replacing the lens, do not over-tighten the screws and crack the plastic.

16 Flashing indicator lamp: replacement of bulbs

1   The indicators are fitted with 21 watt bulbs at the front, or in the case of American models with a 23/8 watt bulb also fitted to the tail/stop lamp assembly. The 8 watt element illuminates as a

position indicator. The rear indicators have a 23 watt bulb.

2   To replace a bulb, remove the plastic lens cover by withdrawing the two retaining crosshead screws. Push the bulb in, turn it to the left and withdraw. Note it is important to use a correctly rated bulb otherwise the flashing rate will be altered. Note that bulbs with a double rating on the front indicators have offset fixing pins to prevent incorrect positioning of the bulbs in the holder.

17 Flasher relay unit: location and replacement

1   The flasher unit is located in the left-hand compartment of the dummy fuel tank, and is rubber mounted on the panel to protect it from vibration.

2   If the unit malfunctions a replacement must be obtained as repair is impracticable.

3   Handle the flasher unit carefully, as it will be damaged if dropped.

14.3a Headlamp horizontal beam is adjusted by screw ...

14.3b ... connected to pivoted reflector unit

15.1a Stop/tail lamp lens cover is held by two screws

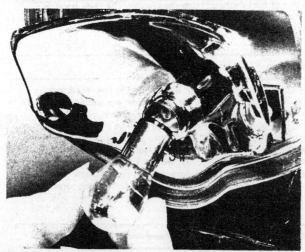

15.1b Bulb is an offset pin 'bayonet' fixed type

16.1a Flashing indicator cover is retained by two screws

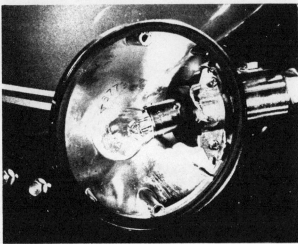

16.1b Bulb is bayonet fixed (offset pin on front indicators)

16.2 Flasher audible warning unit

18 Instrument head and console lamps: replacement

1   The warning bulbs fitted to the instrument heads and the warning console are of 3 - 4 watt rating and are of the bayonet fixing type.
2   The speedometer and tachometer bulbs are retained in the underside of the instruments by rubber push fit holders. Access to the console warning bulbs can be made by removing the upper panel, which is retained by four long crosshead screws.

19 Horn: removal and renovation

1   The horn is located on the right-hand side of the machine and is attached to the frame to the rear of the steering head lug by two bolts. Remove the bolts and disconnect the 'spade' connectors from the rear of the horn to detach the unit from the machine.
2   Some horns are fitted with an adjuster screw at the rear, the rotation of which will alter the pitch of the note produced. If a horn fails it should be renewed as repairs as impracticable.

20 Temperature gauge and sensor: testing

1   If the temperature gauge appears to be faulty first remove the sensor switch from the thermostat housing and test if as follows.
2   Drain the coolant so that leakage will not occur when the sensor switch is removed. Suspend the switch in a pan of water so that the sensor tip is below the water level. Place a thermometer in the water so that the temperature of the water can be taken.
3   Heat the water slowly and take resistance readings with an ohmeter, which should be as follows for the different temperatures.

$60^{\circ}C$	$(140^{\circ}F)$	104.0 ohms
$85^{\circ}C$	$(185^{\circ}F)$	43.9 ohms
$110^{\circ}C$	$(232^{\circ}F)$	20.3 ohms
$120^{\circ}C$	$(250^{\circ}F)$	16.1 ohms

If the sensor does not react as specified, the temperature sensor switch must be renewed.
4   If the temperature gauge is found to be faulty by elimination, after testing the switch, the gauge complete with the speedometer in which it is incorporated, must be renewed.
5   When refitting the switch, apply sealing compound to the threads.

21 Electric fan motor and switch: testing

1   If it is found that the fan does not switch on automatically even when the engine is excessively warm, disconnect the fan motor and test it independently from the switch.
2   Using a 12 volt power source connect the positive lead to the BLUE terminal and the negative lead to the BLACK terminal. If the motor functions correctly renew the sensor switch, which can be removed from the thermostat housing after lowering the coolant level by draining.

22 Fuel gauge and float switch: testing

1   Because the fuel is contained in a tank below the dualseat and therefore verification of the amount of fuel is difficult, a fuel gauge is incorporated in the system, mounted in the centre panel of the dummy fuel tank.
2   If the fuel gauge malfunctions, carry out the following test to isolate where the trouble lies. Remove the dualseat after removing the two retaining bolts. Pull the two leads off the float switch situation in top of the tank. Turn on the ignition and touch the two leads together. The needle on the fuel gauge should deflect to the 'full' position. If the gauge functions satisfactorily, suspect the float switch. The switch requires a special tool for removal, and it is therefore advised that the machine be returned to a Honda service agent for switch replacement.

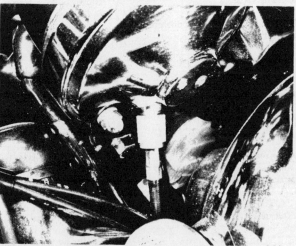

18.1 Instrument heads are retained by acorn nuts

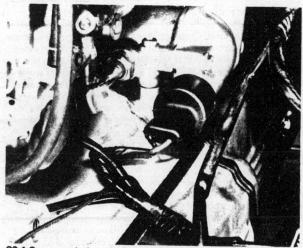

20.1 Sensor switches are fitted to thermostat housing

22.1 Fuel gauge leads have spade connectors

23 Fault diagnosis: electrical system

Symptom	Cause	Remedy
Complete electrical failure	Blown fuse(s)	Check wiring and electrical components for short circuit before fitting new fuse
	Isolated battery	Check battery connections, also whether connections show signs of corrosion
Dim lights, horn and starter inoperative	Discharged battery	Remove battery and charge with battery charger. Check generator output and voltage regulator settings
Constantly blowing bulbs	Vibration or poor earth connection	Check security of bulb holders. Check earth return connections
Parking lights dim rapidly	Battery will not hold charge	Renew battery at earliest opportunity
Tail lamp fails	Blown bulb or fuse	Renew
Headlamp fails	Blown bulb or fuse	Renew
Flashing indicators do not operate, or flash fast or slow	Blown bulb	Renew bulb
	Damaged flasher unit	Renew flasher unit
Horn inoperative or weak	Faulty switch	Check switch
	Out of adjustment	Re-adjust
Incorrect charging	Faulty coil	Check
	Faulty rectifier	Check
	Faulty regulator	Check and adjust
	Wiring fault	Check
Over or under-charging	As above, or faulty battery	Check
Starter motor sluggish	Worn brushes	Remove starter motor and renew brushes
	Dirty commutator	Clean
Starter motor does not turn	Machine in gear	Disengage clutch
	Emergency switch in OFF position	Turn on
	Faulty switches or wiring	Check continuity
	Battery flat	Charge

149

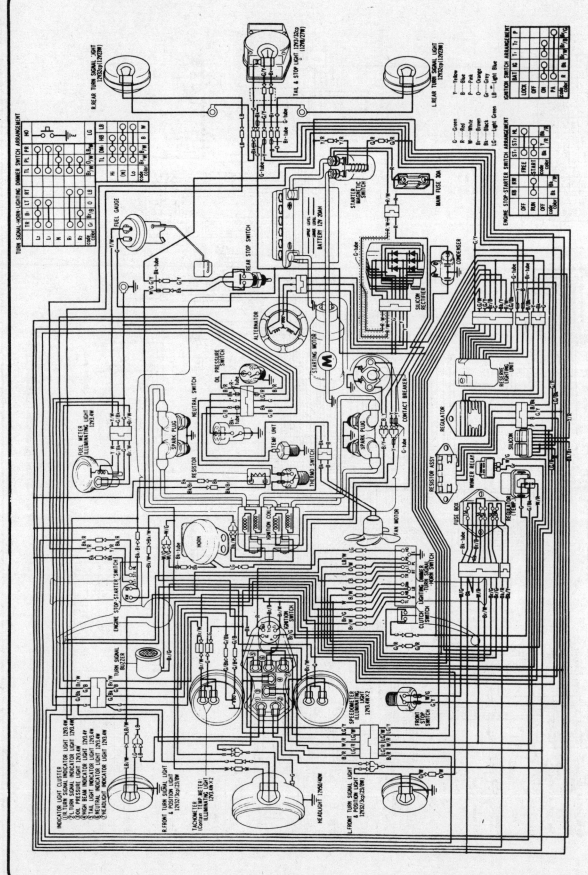

Fig. 7.6. Wiring diagram USA Type

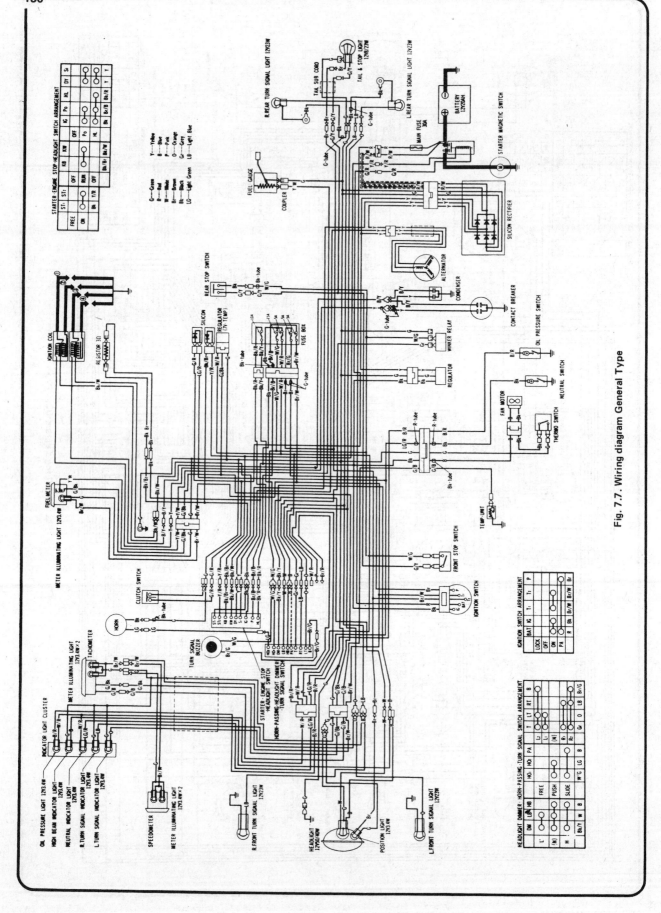

Fig. 7.7. Wiring diagram General Type

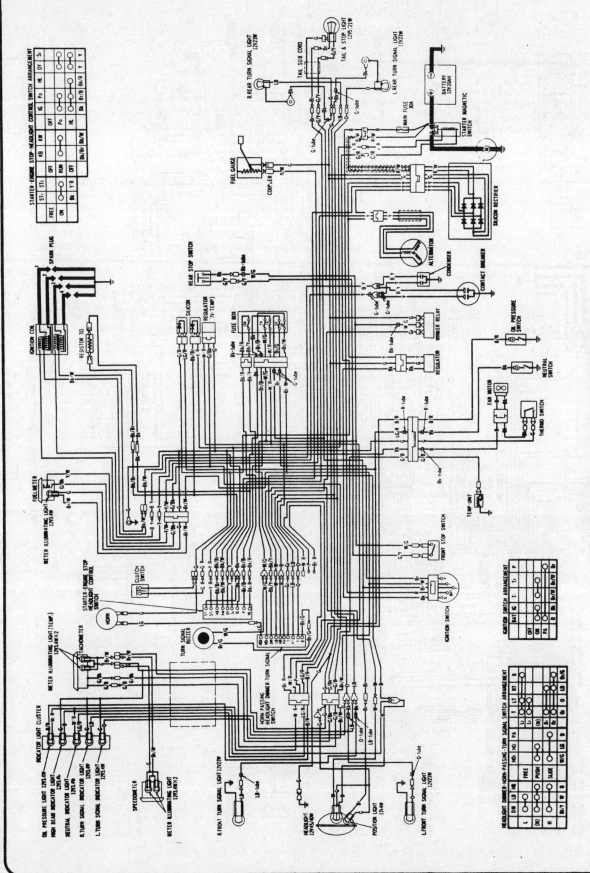

Fig. 7.8. Wiring diagram UK and European Types

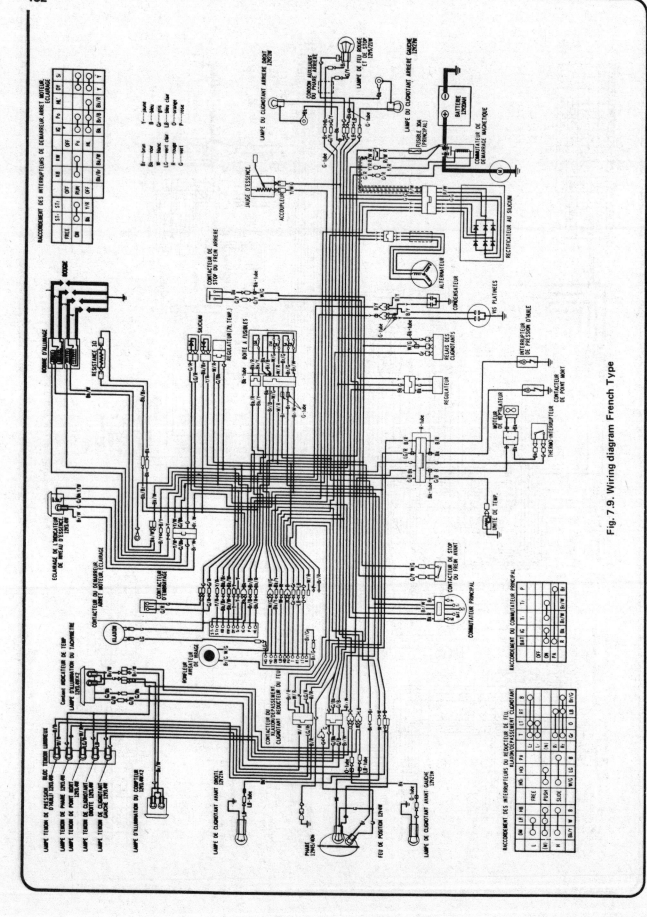

Fig. 7.9. Wiring diagram French Type

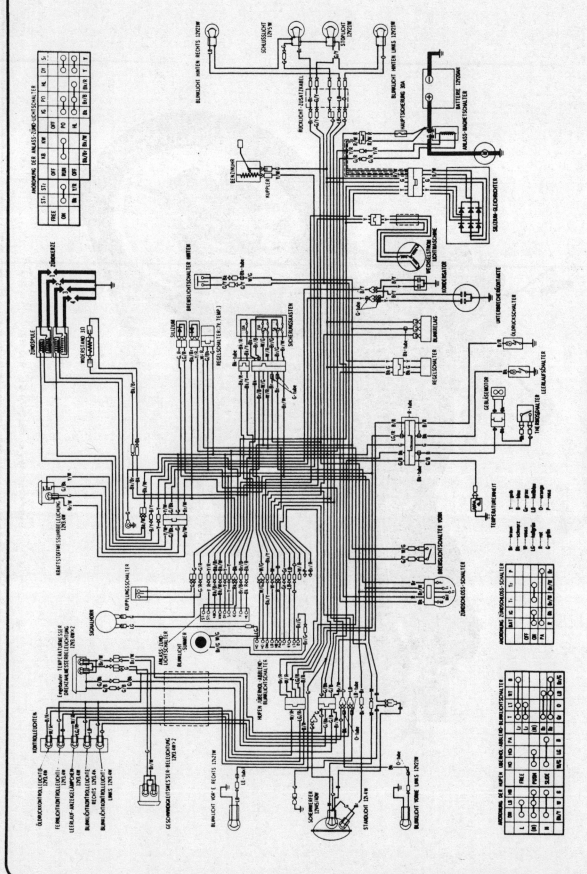

Fig. 7.10. Wiring diagram German Type

Right-hand view of the 1979 GL 1000 KZ

Chapter 8 Honda GL 1000 KZ (UK) and 1978-79 US models

Contents

Specifications

Specifications for the KZ (UK) model and the 1978 — 79 US models are as stated in the main text of this manual, except where listed below:

Specifications relating to Chapter 1

Engine
Valve timing:

Inlet opens	5° BTDC
Inlet closes	35° ABDC
Exhaust opens	40° BBDC
Exhaust closes	5° ATDC

Pistons

Gudgeon pin hole diameter	17.01 — 17.016 mm (0.6697 — 0.6699 in)
Gudgeon pin diameter	17.0 ± 0.003 mm (0.6697 ± 0.0001 in)

Cylinders

Bore taper service limit	0.05 mm (0.002 in)
Maximum ID variation between cylinders	0.10 mm (0.004 in)

Camshafts

Right and left bearing service limit	27.05 mm (1.064 in)
Centre bearing service limit	25.05 mm (0.986 in)
Lobe height:	
Inlet	36.72 — 36.88 mm (1.445 — 1.452 in)
Service limit	36.5 mm (1.437 in)
Exhaust	36.72 — 36.88 mm (1.445 — 1.452 in)
Service limit	36.5 mm (1.437 in)

Gearbox
Gear change drum groove width ... ...   7.05 – 7.15 mm (0.278 – 0.282 in)

Specifications relating to Chapter 3

Carburettors

	1979 KZ (UK)	1978 (US)	1978 EC	1979 (US)
Make	Keihin	Keihin	Keihin	Keihin
Type	770A	769A	771A	771A
Bore diameter	31 mm	31 mm	31 mm	31 mm
Primary main jet	60	60	60	60
Secondary main jet	120	120	120	120
Pilot fuel jet	35	35	35	35
Primary air jet	120	120	140	140
Secondary air jet	60	60	60	60
Pilot air jet	120	120	130	130
Pilot screw (no. of turns out from fully in)	1½	1½	2	2
Float level	21 mm (0.827 in)	21 mm (0.827 in)	21 mm (0.827 in)	21 mm (0.827 in)

Note the EC suffix denotes US emission controlled motorcycles manufactured after 31.12.77

Specifications relating to Chapter 4

Ignition timing ... ...   10° BTDC at 950 ± 100 rpm

Spark plug
Type:
   USA ... ...   NGK D8EA or ND X 24 ES – U
   UK ... ...   NGK DR8ES – L or equivalent

Specifications relating to Chapter 5

Front forks
Spring free length:
   USA ... ...   525 mm (20.67 in)
   UK ... ...   519 mm (20.433 in)
Service limit:
   USA ... ...   500 mm (19.69 in)
   UK ... ...   495 mm (19.488 in)

Rear suspension unit
Spring free length:
   USA ... ...   178.2 mm (7.02 in) and 68.5 mm (2.70 in)
   UK ... ...   248.6 mm (9.787 in)
Service limit:
   USA ... ...   174.9 mm (6.89 in) and 67.0 mm (2.64 in)
   UK ... ...   244 mm (9.606 in)

Specifications relating to Chapter 6

Wheels
Type ... ...   Comstar

Brakes
Front:
   Master cylinder bore diameter – 1979 US model ... ...   15.870 – 15.913 mm (0.6248 – 0.6265 in)
   Service limit ... ...   15.925 mm (0.6270 in)
   Master cylinder piston diameter – 1979 US model ... ...   15.827 – 15.854 mm (0.6231 – 0.6242 in)
   Service limit ... ...   15.815 mm (0.6224 in)
   Disc thickness ... ...   4.8 – 5.2 mm (0.189 – 0.205 in)
   Service limit ... ...   4.0 mm (0.157 in)
   Disc runout (max) ... ...   0.3 mm (0.0118 in)
Rear:
   Caliper bore diameter ... ...   42.85 – 42.90 mm (1.687 – 1.689 in)
   Service limit ... ...   42.915 mm (1.690 in)
   Caliper piston diameter ... ...   42.772 – 48.822 mm (1.684 – 1.686 in)
   Service limit ... ...   42.757 mm (1.683 in)
   Disc thickness ... ...   6.9 – 7.1 mm (0.2716 – 0.2795 in)
   Service limit ... ...   6.0 mm (0.2362 in)
   Disc runout (max) ... ...   0.3 mm (0.0118 in)

Specifications relating to Chapter 7

Bulbs	GL1000 KZ (UK)	GL 1000 '78, '79 (US)
Headlight	QH 55/60W	QH 55/60W
Tail/stop light	5/21W	8/27W (X 2 US model)
Position/front indicator	21W	8/23W
Rear indicator	21W	23W
Instruments	3.4W	3.4W
Indicator warning	3.4W	3.4W

Dimensions and weights

	GL1000 KZ	GL1000 '78 (US)	GL1000 '79 (US)
Overall length	2350 mm (92.5 in)	2320 mm (90.8 in)	2320 mm (90.8 in)
Overall width	759 mm (29.9 in)	920 mm (36.2 in)	920 mm (36.2 in)
Height	1150 mm (45.3 in)	1265 mm (49.8 in)	1255 mm (48.6 in)
Wheelbase	1545 mm (60.8 in)	1545 mm (60.8 in)	1545 mm (60.8 in)
Ground clearance	140 mm (5.5 in)	140 mm (5.5 in)	140 mm (5.5 in)
Weight (dry)	273 kg (601 lb)	273 kg (601 lb)	274 kg (604 lb)
Engine weight	106 kg (234 lb)	107 kg (236 lb)	107 kg (236 lb)

1 Introduction

Although the Honda GL 1000 model types covered in this Chapter are similar in a great many respects to the models covered in the first seven Chapters of this manual, reference should always be made to this Chapter first in view of the need to follow a modified procedure or use different settings, when certain components have to be removed and replaced. Where no information is given in this Chapter, it can be assumed that the procedure is essentially the same as that described for the earlier models.

The GL 1000 models covered in the main text are those imported into the UK between March 1975 and April 1979, the K1 and K2 models, and those imported into the USA between the years 1975 and 1977. The models covered in this Chapter are the KZ model, imported into the UK in April 1979 and discontinued in September of the same year, and the model imported into the USA for the final year of production, 1978 — 79.

Although many minor improvements were made to the models produced between 1975 and 1977, the 1978 — 79 model incorporates many major improvements introduced to give less unsprung weight, better braking, greater rider comfort and an increase in low- and mid-range engine performance. These improvements include the introduction of steel-spoked Comstar wheels, the fitting of front discs and brake calipers of the same design as those fitted to the 750 F2 model, and the uprating of the front forks to give more travel and a changed damping rate. Engine performance has been changed by the fitting of carburettors with a smaller bore diameter, a change in valve timing and a change in ignition timing.

The one remaining major change is the fitting of a redesigned exhaust system, designed to allow better accessibility to the clutch and to emit a slightly louder exhaust note.

2 Air filter and crankcase breather reservoir: removal and cleaning

1 To remove and clean the air filter element, follow the procedure given in Section 15 of Chapter 3. Detail changes to the filter assembly are shown in the figure accompanying this text.

2 US models are fitted with a crankcase breather reservoir which must be removed and emptied at the recommended interval of 3750 miles (6000 km) by carrying out the following procedure. Release the clip securing the drain tube to the reservoir and pull the tube off the reservoir stub. Detach the reservoir from its mounting clip by unscrewing the clip retaining bolt. Remove the flow restrictor from the reservoir stub and empty the contents of the reservoir into a container. Refit the reservoir using a reversal of the above procedure.

3 Note that if the machine is ridden in a humid climate, in rainy conditions, or at full throttle for long periods of time, then the transparent section of the drain tube should be inspected to check the deposit level at more frequent intervals and the reservoir removed and drained accordingly.

3 Exhaust system: removal and refitting

1 The exhaust system differs from the system fitted to earlier models in that it comprises two separate system halves, each half consisting of a two-into-one exhaust pipe clamped to a silencer. The two system halves are interconnected by a balancer pipe, the pipe being clamped to a stub on each silencer.

2 To remove the complete exhaust system from the machine, proceed as follows, bearing in mind that each system half may be removed independent of the other should the need arise.

Removal of the complete system will be more easily achieved with the aid of an assistant because the assembly weighs a considerable amount and will need to be properly supported throughout the removal procedure.

3   Commence removal by unscrewing and removing the two nuts from each of the four exhaust pipe to cylinder head connections. Remove the bolt from one of the balancer pipe to silencer retaining clamps and check that the clamp is free. Loosen the nut and bolt which retains each silencer to each pillion footrest bracket and remove each nut. With one person each side of the machine supporting the exhaust assembly, withdraw the mounting bolt from each pillion footrest bracket and carefully lower the complete assembly away from its mounting points. Once clear, the balancer pipe may be separated at its connection and the two separate exhaust system halves lifted clear of the machine. Under no circumstances should the exhaust assembly be allowed to hang unsupported from its cylinder head mounting points because the weight of the system will place an unacceptable strain on the cylinder head studs.

4   Once removed from the machine, each exhaust pipe may be separated from its silencer by removing the clamp retaining bolt and pulling the pipe out of its location in the silencer. The heat guards may be removed from the pipes and silencers by unscrewing the retaining bolts and screws. Care should be taken when loosening these bolts and screws to use a spanner or screwdriver of the correct size because the extreme changes in temperature to which the threads of the bolts and screws have been subjected will have caused them to have become partially seized and some force will be required to initiate removal. A tool of the incorrect size will cause the bolt or screw head to become damaged, making removal of the item very difficult. When reassembling the system lightly smear the threads of the heat guard retaining bolts and screws with graphite grease to reduce the likelihood of seizure on the next occasion of removal. Do not omit to refit the spacer band to the end of the exhaust pipe before inserting the pipe into the silencer and tightening the clamp retaining bolt finger-tight.

5   As with removal, refitting of the exhaust system requires the aid of an assistant. Fit a new gasket into each exhaust port, distorting them very slightly if necessary so that they become oval and remain in the ports. With one person each side of the machine, push the two halves of the system together at the balancer pipe connection. The complete assembly may then be lifted up to locate on the cylinder head studs and align with the rear mounting points. Refit the bolt and nut to each rear mounting and the two nuts to each of the four cylinder head manifold connections. Tighten the manifold nuts, evenly, to avoid distortion of the pipe flanges. Tighten both rear mounting nuts followed by the balancer pipe clamp bolt and the two exhaust pipe to silencer clamp bolts.

6   On no account run the machine with the exhaust baffles removed, or with a quite different type of silencer fitted. The standard production silencers have been designed to give the best possible performance, whilst subduing the exhaust note to an acceptable level. Although a modified exhaust system, or one without baffles may give the illusion of greater speed as a result of the changed exhaust note, the chances are that performance will have suffered accordingly.

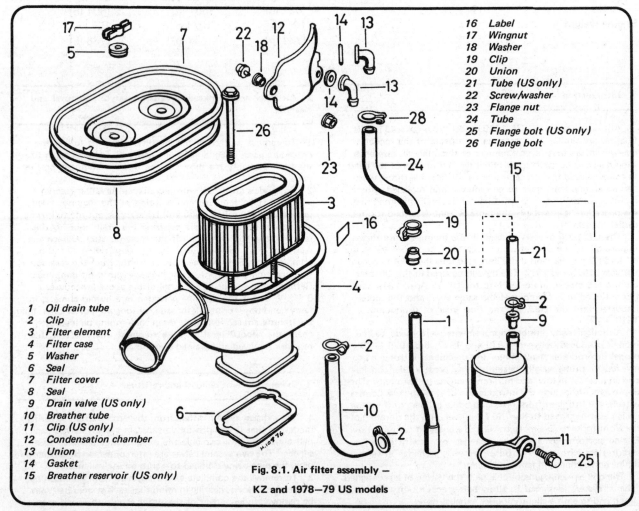

16   Label
17   Wingnut
18   Washer
19   Clip
20   Union
21   Tube (US only)
22   Screw/washer
23   Flange nut
24   Tube
25   Flange bolt (US only)
26   Flange bolt

1   Oil drain tube
2   Clip
3   Filter element
4   Filter case
5   Washer
6   Seal
7   Filter cover
8   Seal
9   Drain valve (US only)
10   Breather tube
11   Clip (US only)
12   Condensation chamber
13   Union
14   Gasket
15   Breather reservoir (US only)

**Fig. 8.1. Air filter assembly —
KZ and 1978—79 US models**

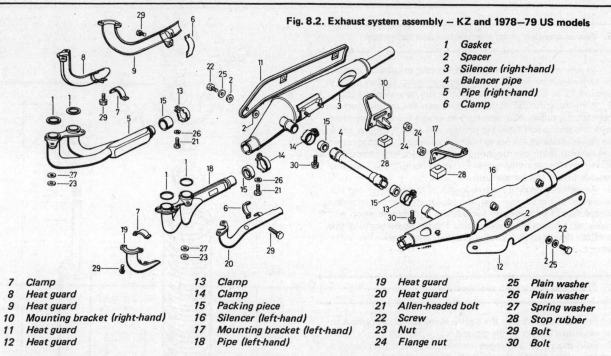

Fig. 8.2. Exhaust system assembly — KZ and 1978—79 US models

1	Gasket
2	Spacer
3	Silencer (right-hand)
4	Balancer pipe
5	Pipe (right-hand)
6	Clamp

7	Clamp	13	Clamp	19	Heat guard	25	Plain washer
8	Heat guard	14	Clamp	20	Heat guard	26	Plain washer
9	Heat guard	15	Packing piece	21	Allen-headed bolt	27	Spring washer
10	Mounting bracket (right-hand)	16	Silencer (left-hand)	22	Screw	28	Stop rubber
11	Heat guard	17	Mounting bracket (left-hand)	23	Nut	29	Bolt
12	Heat guard	18	Pipe (left-hand)	24	Flange nut	30	Bolt

4 Steering head bearings: removal, examination, renovation, fitting and adjustment

1   The steering head bearings fitted to the later models are of the taper roller variety, and are unlikely to give rise to problems in the normal life of the machine, although the manufacturer recommends examination, lubrication and adjustment of these components at 12000 mile (20 000 km) intervals. Access to the bearings is gained by following the procedure detailed in Section 2 of Chapter 5.

2   It will be noted that the bearings are effectively in two parts; the outer races, which will remain in the steering head tube, and the inner race, cage and rollers, which will come away as the lower yoke is removed. Of the latter, the lower bearing will probably be firmly attached to the steering stem, whilst the upper bearing will lift away quite easily. It is normally possible to lever the lower bearing off the steering stem, but it may prove necessary to employ a bearing extractor in stubborn cases.

3   The outer races can be removed with the aid of a long drift, passed through the steering head tube, new races being fitted by judicious use of a tubular drift, such as a large socket or similar. Before any decision is made to remove the outer races, they should be cleaned and checked as described below.

4   Wash out the bearings with clean petrol to remove all traces of old grease or dirt. Check the faces of the rollers and the outer races for signs of wear or pitting, both of which are unlikely unless the machine has been neglected in the past. If damaged, the bearings must be renewed.

5   Note that the bearing and the bearing race must be renewed as a set. Ensure when removing and fitting the bearing races that they leave and enter the steering head tube squarely and are properly seated in the tube. The same rule applies when fitting the lower bearing to the steering stem. Refer to the figure accompanying this text when fitting the bearings and assembling the steering head, noting the position of the dust seal and grease retainer. Renew the dust seal if the lower bearing is removed.

6   When assembling the steering head, pack each bearing with the recommended grease prior to installation. Offer up the lower yoke to the steering head tube and retain it in position by fitting the adjuster nut finger-tight, after having first fitted the upper bearing. Adjustment of the bearings requires a socket-type peg spanner so that the slotted adjuster nut can be set to the prescribed torque loading. This requires the use of the special Honda steering stem socket (Part No 07916-3710100) or a home-made equivalent. A piece of tubing can be filed to fit the nut and then welded to a damaged socket to improvise.

7   Tighten the adjuster nut to a torque of 3 — 4 kgf m (22 — 28 lbf ft) to seat the bearings. Loosen the adjuster nut and retighten it to a torque of 1.5 — 1.7 kgf m (11 — 13 lbf ft). Check that the lower yoke turns freely from side to side, with no resistance.

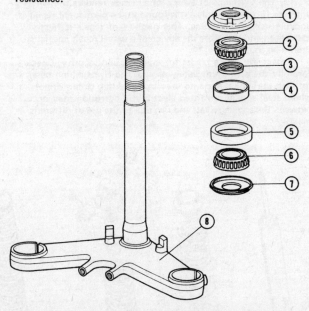

Fig. 8.3. Steering head bearing assembly — KZ and 1978—79 US models

1	Top nut	5	Lower bearing outer race
2	Upper bearing		
3	Grease retainer	6	Lower bearing
4	Upper bearing outer race	7	Dust seal
		8	Steering stem

5   Rear suspension units: examination and renovation

1   Examination and renovation of the rear suspension units may
be carried out by following a procedure similar to that listed in
Section 11 of Chapter 5, whilst noting the following points.
2   With the spring(s) compressed, loosen the locknut located
beneath the upper mounting lug and unscrew the lug together
with the spring seat from the damper rod end. A tommy bar may
be passed through the lug to hold it in position whilst loosening
the locknut. With the mounting lug and spring seat removed, the
spring may be carefully released from its compressed state and
removed from the damper unit.
3   As well as checking the free length of the spring(s) and the
operation of the damper unit, the rubber stopper and mounting
pieces should be inspected for damage and deterioration and
renewed if necessary. Refer to the figures accompanying this
text for details of the various types of rear suspension unit
assemblies.

6   Front and rear wheel: examination

1   Place the machine on the centre stand so that the wheel to
be examined is clear of the ground.
2   Spin the wheel and check for rim alignment by placing a
pointer close to the rim edge. If the total radial or axial
alignment variation is greater than 2.00 mm (0.08 in) the
manufacturers recommend that the wheel is renewed. This
policy is, however, a counsel of perfection and in practice a
larger runout may not affect the handling properties excessively.
3   Although Honda do not offer any form of wheel rebuilding
facility, a number of private engineering firms offer this service.
It should be noted however, that Honda do not approve of this
course of action.
4   Check the rim for localised damage in the form of dents or
cracks. The existence of even a small crack renders the wheel
unfit for further use unless it is found that a permanent repair is
possible using arc-welding. This method of repair is highly
specialised and therefore the advice of a wheel repair specialist
should be sought.
5   Inspect the spoke blades for cracking and security. Check
carefully the area immediately around the rivets which pass
through the spokes and into the rim. In certain circumstances
where steel spokes are fitted electrolytic corrosion may occur
between the spokes, rivets and rim due to the use of different
metals.

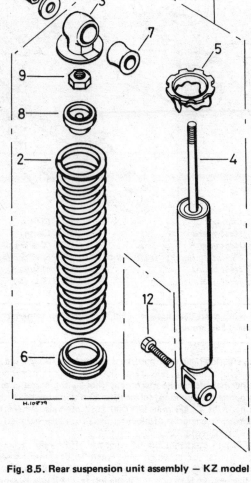

Fig. 8.5. Rear suspension unit assembly — KZ model

1	Suspension unit assembly	7	Mounting rubber
2	Spring	8	Rubber stopper
3	Spring seat/mounting lug	9	Hexagon nut
4	Damper unit	10	Mounting nut
5	Adjuster plate	11	Washer
6	Spring seat	12	Bolt

Fig. 8.4. Rear suspension unit assembly — 1978—79 US models

1	Suspension unit assembly
2	Spring
3	Spring
4	Spring seat/mounting lug
5	Spring seat
6	Damper unit
7	Adjuster plate
8	Spring seat
9	Spring seat
10	Mounting rubber
11	Rubber stopper
12	Hexagon nut
13	Mounting nut
14	Washer
15	Bolt

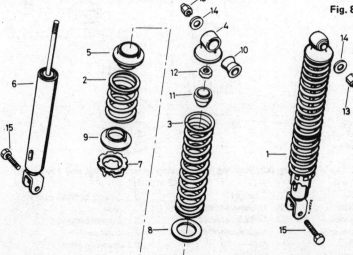

7   Front wheel: removal and refitting

1   With the front wheel supported well clear of the ground, remove the cross head screw which retains the speedometer cable to its drive gearbox, on the left-hand side of the hub. Pull the cable out and refit the screw, to prevent loss.

2   Remove the two bolts holding one of the caliper support brackets to the fork leg and lift the caliper and bracket assembly off the disc. Support the weight of the caliper with a length of string or wire attached to the frame or engine.

3   Slacken and remove the clamp nuts at the base of each fork leg. With the clamps released, the wheel will drop free and can be manoeuvred clear of the forks and mudguard.

4   Do not operate the front brake lever while the wheel is removed since fluid pressure may displace the pistons and cause leakage. Additionally, the distance between the pads will be reduced, making refitting of the brake discs more difficult.

5   Refit the wheel by reversing the dismantling procedure. Do not omit the spacer which is a push fit in the oil seal on the right-hand side of the wheel or the speedometer gearbox which is a push fit on the left-hand side. Ensure that the speedometer drive dogs engage with the notches in the gearbox drive sleeve. Lift the wheel into position whilst ensuring that the speedometer gearbox is positioned correctly. Fit the clamps to the base of each fork leg; the arrow marked on each clamp must face forward. Fit the clamp retaining nuts and washers finger-tight.

6   Carefully lower the disconnected brake caliper assembly over the disc to avoid damage to the pads and fit and tighten the two securing bolts to the specified torque loading. Tighten the nuts of the fork leg clamp on the left-hand side to the specified torque loading, starting with the forward nuts.

7   Before tightening the right-hand clamp securing nuts, it is first necessary to determine that the clearance between the outside surface of the right-hand disc and the rear of the caliper support bracket is correct. If this clearance is not correct, damage to the disc is likely to occur resulting in impaired braking efficiency or worse. With a feeler gauge of 0.7 mm (0.028 in) thickness, measure the clearance. If the gauge inserts easily into the space between the disc and support bracket, torque load the clamp securing nuts to the specified figure, starting with the forward nut.

8   If the feeler gauge will not insert into the space, grasp the right-hand fork lower leg and pull it outwards until the gauge can be inserted. Tighten the clamp securing nuts as stated above and withdraw the feeler gauge. Check that the clearance between the disc surface and the other three corners of the caliper support bracket is also 0.7 mm (0.028 in). Spin the wheel to ensure that it revolves freely and check the brake operation. Check that all nuts and bolts are fully tightened. If the clearance between the disc and pads is incorrect pump the front brake lever several times to adjust. Finally, reconnect the speedometer cable to the speedometer gearbox.

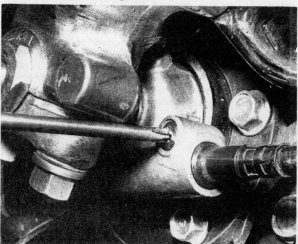

7.1 Remove the retaining screw to release the speedometer cable

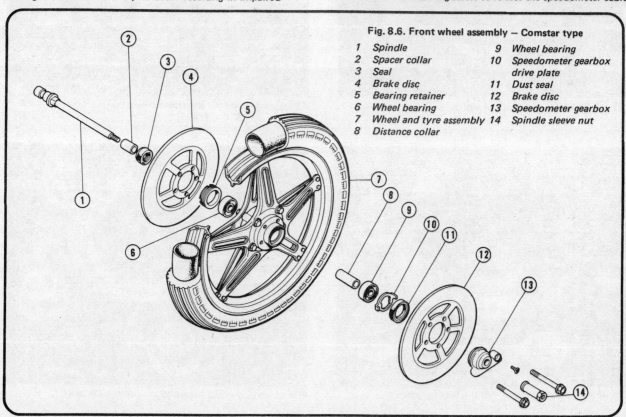

Fig. 8.6. Front wheel assembly — Comstar type

1	Spindle	9	Wheel bearing
2	Spacer collar	10	Speedometer gearbox
3	Seal		drive plate
4	Brake disc	11	Dust seal
5	Bearing retainer	12	Brake disc
6	Wheel bearing	13	Speedometer gearbox
7	Wheel and tyre assembly	14	Spindle sleeve nut
8	Distance collar		

8 Front wheel bearings: removal, examination and fitting

1 With the wheel removed from the machine, unscrew and remove the sleeve nut from the wheel spindle. Pull the speed-ometer gearbox from its location in the left-hand side of the wheel hub and remove the wheel spindle, followed by the spacer contained within the right-hand dust seal. The wheel bearings are of the ball journal type and non-adjustable. There are two bearings and two dust seals, the two bearings being interposed by a distance collar in the centre of the hub. To avoid damage occurring to the brake discs during removal of the bearings, it is recommended that they be removed by following the procedure given in Section 13.

2 Note that a threaded retainer is fitted to the right-hand side of the hub, and this must be removed before the bearings can be released. A suitable peg spanner should be fabricated unless the Honda tool, No 07710-0010200, is available (see Section 15, paragraph 2). Do not resort to using a punch to loosen the retainer; this will only result in damage. The retainer will be staked in position and will require firm, even pressure to release it. Note also that new bearings and seals should be fitted whenever the old items are removed, so check carefully for wear before dismantling commences.

8.6a Locate the speedometer drive plate in the wheel hub ...

8.6b ... followed by the new dust seal ...

8.6c ... and locate the speedometer gearbox

8.7a Insert the new dust seal into the bearing retainer ...

8.7b ... and insert the spacer collar through the seal

3   The left-hand bearing should be drifted out first, from the right-hand side of the wheel. Use a long drift against the inner face of the inner race. It may be necessary to knock the collar to one side so that purchase can be made against the race. Work round in a circle to keep the bearing square in the housing. The dust seal and speedometer drive dog plate will be pushed out as the bearing is displaced. After removal of the bearing, take out the distance collar and then drift out the right-hand bearing and dust seal in a similar manner, from inside the hub.

4   Remove all the old grease from the hub and bearings, wash the bearings in petrol, and dry them thoroughly. Check the bearings for roughness by spinning them whilst holding the inner track with one hand and rotating the outer track with the other. If there is the slightest sign of roughness renew them.

5   Before driving the bearings back into the hub, pack the hub with new grease and also grease the bearings. Use a tubular drift of the same diameter as the outer race of each bearing to drive the bearings back into the wheel hub. Ensure that the bearings enter the hub squarely and are fitted with the closed side facing outwards. Do not omit to refit the distance collar before fitting the second bearing.

6   Locate the speedometer drive plate in the wheel hub, ensuring that the tab of the plate fits in the slot in the hub casting. Fit the new dust seal over the drive plate. Screw the bearing retainer into position, having obtained a new replacement if there were signs of wear or damage to the threads Tighten the retainer firmly, then secure it in this position by staking at the junction of the retainer and the wheel hub. Refit the brake discs, tightening the retaining bolt nuts evenly and in a diagonal sequence.

7   Refit the spacer into the right-hand dust seal and push the wheel spindle into position in the wheel after first having smeared its bearing surfaces lightly with grease. Lubricate the speedometer gearbox with the correct type of grease and push it into position, aligning the tangs on the gearbox with the notches in the drive plate. Fit and tighten the wheel spindle sleeve nut to the specified torque loading.

9   Front disc brake assembly: examination and brake pad renewal

1   Check the front brake master cylinder, hoses and caliper units for signs of leakage. Pay particular attention to the condition of the hoses, which should be renewed without question if there are signs of cracking, splitting or other exterior damage. Check the hydraulic fluid level by referring to the upper and lower level lines visible on the exterior of the translucent reservoir body.

2   Replenish the reservoir after removing the cap on the brake fluid reservoir and lifting out the diaphragm plate. The condition of the fluid is one of the maintenance tasks which should **never be neglected.** If the fluid is below the lower level mark, brake fluid of the correct specification must be added. **Never** use engine oil or any fluid other than that recommended. Other fluids have unsatisfactory characteristics and will rapidly destroy the seals.

3   The two sets of brake pads should be inspected for wear. Each has a red groove, which marks the wear limit of the friction material. When this limit is reached, both pads in the set must be renewed, even if only one has reached the wear mark. A small inspection window, closed by a plastic cap, is provided in the top of each caliper unit so that examination of pad condition may be carried out easily.

4   If the brake action becomes spongy, or if any part of the hydraulic system is dismantled (such as when a hose has been renewed) it is necessary to bleed the system in order to remove all traces of air. Follow the procedure in Section 12 of this Chapter.

5   To gain access to the pads for renewal, the caliper assembly being attended to must be partially dismantled as follows; removal of the wheel is not required. Unscrew the two bolts that pass into the caliper casing. Carefully ease the casing up, leaving the pads in position either side of the disc and supported by the caliper mounting bracket. The pads can be lifted from place, one at a time.

6   Refit the new pads and the caliper by reversing the dismantling procedure. The caliper piston should be pushed inwards slightly so that there is sufficient clearance between the brake pads to allow the caliper to fit over the disc. Do not omit the anti-chatter shim which should be fitted on the rear face of the piston side pad with the arrow pointing forwards (facing the direction of wheel rotation). It is recommended that the outer periphery of the outer (piston) pad is lightly coated with disc brake assembly grease (silicone grease). Use the grease sparingly and ensure that grease **does not** come into contact with the friction surface of the pad.

7   Before refitting the caliper casing, check the condition of the dust covers around the slider spindles and of the spindles themselves. Check also the condition of the caliper piston seal and boot. If the condition of any of these components is seen to be doubtful, refer to the following Section for further information. With the caliper casing refitted, fit and tighten the two securing bolts to the recommended torque loading.

9.5a Detach the caliper from its mounting bracket ...

9.5b ... to allow access to the brake pads

9.6 Refit the shim with the arrow pointing in the direction of wheel rotation

10 Front brake calipers: examination and overhaul

1 It is recommended that the two caliper units are removed and overhauled separately, to prevent the accidental transposition of identical components. Note that any work on the hydraulic system must be undertaken in ultra-clean conditions. Particles of dirt will score the working parts and cause early failure. Select a suitable receptacle into which may be drained the hydraulic fluid. Remove the banjo bolt holding the hydraulic hose at the caliper and allow the fluid to drain. Take great care not to allow hydraulic fluid to spill onto paintwork; it is a very effective paint stripper. Hydraulic fluid will also damage rubber and plastic components.

2 Remove the caliper from the fork leg and displace the brake pads as described in the preceding Section. Withdraw the two slider spindles and rubber boots from the support bracket.

3 Displace the circlip which holds the piston boot in position and then prise out the piston boot, using a small screwdriver, taking care not to scratch the surface of the cylinder bore. The piston can be displaced most easily by applying an air jet to the hydraulic fluid feed orifice. Be prepared to catch the piston as it falls free. Displace the annular piston seal from the cylinder bore groove, again using the flat of a small screwdriver and taking care not to scratch the surface of the cylinder bore.

4 Clean the caliper components thoroughly in trichlorethylene or in hydraulic brake fluid. **CAUTION:** Never use petrol for cleaning hydraulic brake parts otherwise the rubber components will be damaged. Discard all the rubber components as a matter of course. The replacement cost is relatively small and does not warrant re-use of components vital to safety.

5 Check the piston and caliper cylinder bore for scoring, rusting or pitting. If any of these defects are evident it is unlikely that a good fluid seal can be maintained and for this reason the components should be renewed. Inspect the slider spindles for wear and check their fit in the support bracket. Slack between the spindles and bores may cause brake judder if wear is severe.

6 To assemble the caliper, reverse the removal procedure. When assembling pay attention to the following points. Apply caliper grease (high heat resistant) to the caliper spindles. Apply a generous amount of brake fluid to the inner surface of the cylinder and to the periphery of the piston, then reassemble. Do not reassemble the piston with it inclined or twisted. When installing the piston push it slowly into the cylinder while taking care not to damage the piston seal. Apply brake pad grease around the periphery of the moving pad.

7 Refer to Section 10 of Chapter 6 and bleed the brake system after refilling the reservoir with new hydraulic fluid of the correct specification. See also Section 12 of this Chapter. Check for leakage whilst applying the brake lever tightly and repeat the entire servicing procedure for the second brake caliper. Test the operation of the brakes by pushing the machine forward and applying the brake lever. If this test is satisfactory, test run the machine, applying the brakes soon after riding away and at intervals thereafter and noting the level of the hydraulic fluid in the handlebar mounted reservoir to ensure it does not drop. On completion of the test run, recheck the system for signs of leakage and check the disturbed connections for security.

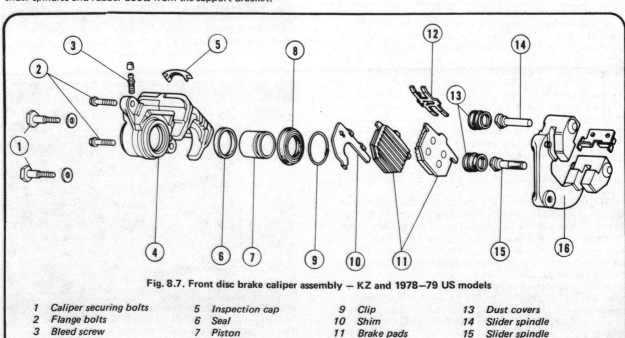

Fig. 8.7. Front disc brake caliper assembly — KZ and 1978—79 US models

1	Caliper securing bolts	5	Inspection cap	9	Clip
2	Flange bolts	6	Seal	10	Shim
3	Bleed screw	7	Piston	11	Brake pads
4	Caliper	8	Boot	12	Spring plate

13	Dust covers
14	Slider spindle
15	Slider spindle
16	Caliper support bracket

11 Front brake master cylinder: removal, examination, renovation and fitting — 1979 US models

1   The master cylinder and hydraulic reservoir take the form of a combined unit mounted on the right-hand side of the handlebars, to which the front brake lever is attached. The master cylinder is actuated by the front brake lever, and applies hydraulic pressure through the system to operate the front brake when the handlebar lever is manipulated. The master cylinder pressurises the hydraulic fluid in the brake pipe which, being incompressible, causes the piston to move in the caliper unit and apply the friction pads to the brake disc. If the master cylinder seals leak, hydraulic pressure will be lost and the braking action rendered much less effective.

2   Before the master cylinder can be removed, the system must be drained. Place a clean container below one caliper unit and attach a plastic tube from the bleed screw on top of the caliper unit to the container. Open the bleed screw one complete turn and drain the system by operating the brake lever until the master cylinder reservoir is empty. Close the bleed screw and remove the pipe.

3   Disconnect the front brake stop lamp switch wire at the push connector. Unscrew the union bolt and disconnect the connection between the master cylinder and brake hose. Tie the hose to a point on the fork assembly and mask its end to prevent the ingress of dirt and moisture into the brake system. Remove the rear view mirror by unscrewing it from the master cylinder body. Unscrew the brake lever retaining nut and bolt and detach the lever from its pivot lugs. Unscrew the two master cylinder fastening bolts and remove the master cylinder body from the handlebars. Empty any surplus fluid from the reservoir.

4   Remove the circlip located beneath the cylinder body boot, followed by the plain washer, piston (with secondary cup), primary cup and the spring. Note that it may be necessary to apply a low pressure air supply to the master cylinder outlet in order to displace the piston. Place the components in a clean container and wash them in new brake fluid. Examine the cylinder bore and piston for scoring. Renew if scored. Examine the brake lever pivot point and the master cylinder pivot lugs for wear, cracks or fractures, and the hose union threads and brake pipe threads for cracks or other signs of deterioration. Remove the reservoir from the cylinder body, by unscrewing and removing the four securing screws. Renew the O-ring attached to the base of the reservoir and inspect the diaphragm for signs of damage or deterioration, renewing it if necessary.

5   When reassembling the master cylinder follow the removal procedure in reverse order. Renew the various seals, lubricating them with silicone grease or hydraulic fluid before they are refitted. Make sure that the primary and secondary cups are fitted the correct way round and that their lips do not turn inside out when being fitted. Check that, once fitted, the circlip is correctly located in its retaining groove.

6   Mount the master cylinder on the handlebars so that the fluid reservoir is horizontal when the motorcycle is on the centre stand with the steering in the straight ahead direction. Note that Honda mark the handlebars and master cylinder securing bracket to ensure correct positioning of the unit. The securing bracket should be fitted with the punch mark facing down and the complete unit positioned so that the punch mark on the handlebars aligns with the joint between the securing bracket and the master cylinder body. Tighten the bracket retaining bolts, upper bolt first. On completion of reassembly, refill the reservoir with new hydraulic fluid of the correct specification and bleed the system.

7   The component parts of the master cylinder assembly and the caliper assemblies may wear or deteriorate in function over a long period of use. It is however, generally difficult to foresee how long each component will work with proper efficiency. From a safety point of view it is best to change all the expendable parts every two years on a machine that has covered a normal mileage.

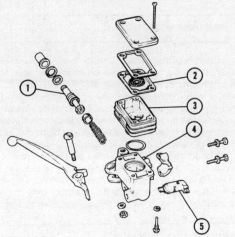

Fig. 8.8. Front brake master cylinder assembly — 1979 US models

1   Piston assembly   4   Master cylinder
2   Diaphragm   5   Brake stop lamp switch
3   Reservoir

12 Bleeding the front brake hydraulic system

1   The procedure for bleeding the front brake hydraulic system remains the same as that given in Section 10 of Chapter 6. It should be noted however, that when operating the brake lever to bleed the system, the bleed valve must be closed just before the distance between the rear face of the end of the lever and the forward face of the throttle twistgrip reaches 15 mm (0.60 in). Do not pull the lever right back to the twistgrip as this will cause piston overtravel resulting in fluid seepage. The lever should not be released until the bleed valve is fully closed otherwise air will be drawn into the system.

13 Brake discs: examination, removal and fitting

1   To carry out examination, removal and fitting of the three brake discs, refer to the procedure given in Section 13 of Chapter 6 whilst checking the Specifications at the beginning of this Chapter for the correct service limits and torque loading figures. Always undo the disc retaining nuts evenly and in a diagonal sequence and tighten them using the same technique; this will avoid any undue strain being placed on the disc with the subsequent risk of distortion occurring.

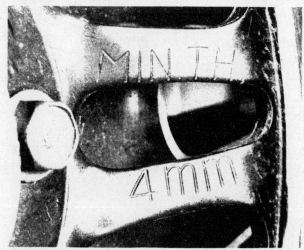

13.1 The wear service limit is marked on each brake disc (front disc shown)

14 Rear wheel: removal and refitting

1   Removal and refitting of the Comstar type of rear wheel may be carried out by following the procedure detailed in Section 6 of Chapter 6, whilst noting the following points.
2   If force is required to dislodge the wheel spindle, the left-hand fork leg of the swinging arm must be properly supported by using a wooden block to apply inward pressure to the end of the fork leg. This is because it is possible to bend the fork leg whilst drifting the spindle out of position.
3   Once the brake caliper is moved clear of the disc, it is worth fitting a wooden wedge between the brake pads to prevent their expulsion should the brake pedal be operated whilst the wheel is removed. When relocating the caliper assembly over the disc, great care must be taken to avoid damage occurring to the pads.

15 Rear wheel bearings: removal, examination and fitting

1   Remove the brake disc, which is retained by six nuts that screw onto studs in the hub. Lift out the final drive splined flange, the pins of which fit into cush drive rubber bushes. If the flange has not been removed for some time, corrosion between the pins and bush sleeves will cause difficulty in removal. A two-legged sprocket puller can be used to aid removal.
2   Remove the bearing retainer from the left-hand side of the wheel hub by using a peg spanner. If the Honda tool No 07710 - 0010100 is not available, this tool may be fabricated from a length of steel bar and two nuts and bolts of the appropriate diameter (see the accompanying photograph). Do not attempt to

drift the retainer loose by using a hammer and punch or similar tools because this will only result in damage occurring. The retainer is staked in position and will require firm even pressure to release it.
3   The left-hand bearing should be drifted out first, from the right-hand side of the wheel. Use a long drift against the inner face of the inner race. It may be necessary to knock the collar to one side so that purchase can be made against the race. Work round in a circle to keep the bearing square in the housing. The dust seal will be pushed out as the bearing is displaced. After removal of the bearing, take out the distance collar and then drift out the right-hand bearing in a similar manner.
4   Remove all the old grease from the hub and bearings, wash the bearings in petrol, and dry them thoroughly. Check the bearings for roughness by spinning them whilst holding the inner track with one hand and rotating the outer track with the other. If there is the slightest sign of roughness renew them.
5   Before driving the bearings back into the hub, pack the hub with new grease and also grease the bearings. Using a tubular drift of the same diameter as the outer race of each bearing, drive the bearings back into the wheel hub. Ensure that each bearing enters the hub squarely and is fitted with the closed side facing outwards. Do not omit to refit the distance collar before fitting the second bearing.
6   Fit the new dust seal over the left-hand bearing and screw the retainer into position, having obtained a new replacement if there were signs of wear or damage to the threads. Tighten the retainer firmly, then secure it in this position by staking at the junction of the retainer and the wheel hub.
7   Refit the brake disc and tighten the nuts evenly and in a diagonal sequence to the specified torque loading. Before refitting the final drive flange, lubricate the damper pins with a multipurpose lithium based grease.

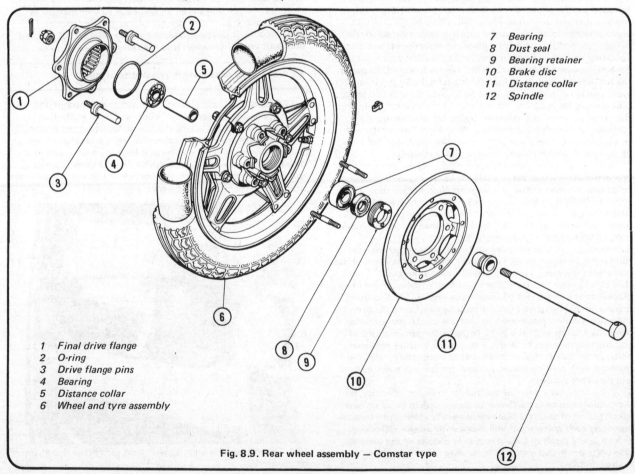

7	Bearing
8	Dust seal
9	Bearing retainer
10	Brake disc
11	Distance collar
12	Spindle

1   Final drive flange
2   O-ring
3   Drive flange pins
4   Bearing
5   Distance collar
6   Wheel and tyre assembly

Fig. 8.9. Rear wheel assembly — Comstar type

15.2 Fabricate a tool with which to remove the bearing retainer

15.3 Use a long drift to remove the wheel bearings

15.5a Do not omit to refit the distance collar ...

15.5b ... before inserting the second bearing into the wheel hub

15.5c Use a tubular drift to drive the bearing into its location

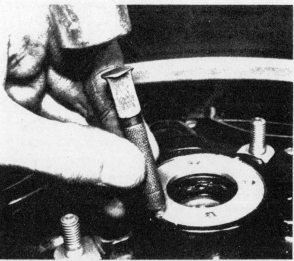
15.6 Stake the bearing retainer to lock it in position

15.7 Lubricate the damper pins before fitting the final drive flange

16 Rear brake: renewing the brake pads and overhauling the caliper unit

1   Although the rear brake caliper assembly is of a slightly different design to that fitted to earlier models, the servicing procedures are as listed in Section 11 of Chapter 6. When carrying out any servicing on the caliper assembly, refer to the service limits given in the Specifications Section of this Chapter and also to the figure accompanying this text.

17 Tail/stop lamp: bulb renewal — 1979 US models

1   The tail lamp assembly fitted to later 1979 US models contains two twin-filament bulbs. To gain access to these bulbs, unscrew and remove the four crosshead screws that retain the plastic lens in position. Separate the lens from the lamp unit, taking care to ensure that the rubber seal is retained in position on the case rim and does not tear or split.
2   The bulbs both have a bayonet fitting with offset pins so that they can be fitted in one position only. Take care not to touch the glass envelope of the new bulb when fitting, use a dry, clean cloth or tissue. To remove a bulb, push it inwards and turn it anti-clockwise so that the bayonet pins disengage. Fitting a bulb is the reversal of this procedure.
3   When refitting the lens, take care not to over-tighten the screws and crack the plastic. Refer to the Specifications at the beginning of this Chapter for bulb wattage ratings and to the figure accompanying this text for details of the tail lamp assembly.

18 Direction indicator lamps: bulb renewal — 1979 US models

1   To renew the bulbs fitted in the rectangular type of indicators fitted to the 1979 US models, follow the procedure described in paragraph 2, Section 16 of Chapter 7, noting that the lenses are retained by three crosshead screws. Refer to the Specifications at the beginning of this Chapter for bulb wattage ratings and to the figure accompanying this text for details of the indicator assembly.

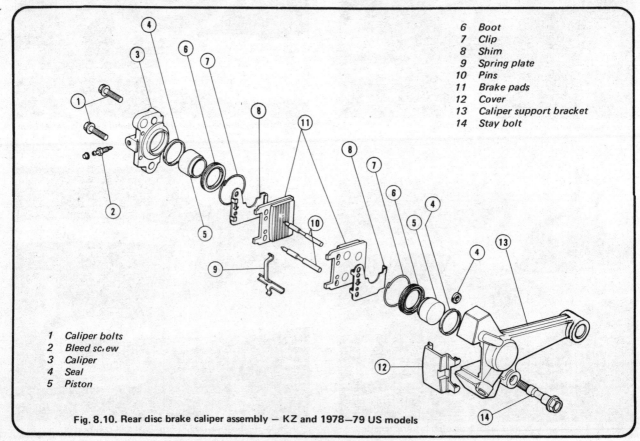

6   Boot
7   Clip
8   Shim
9   Spring plate
10   Pins
11   Brake pads
12   Cover
13   Caliper support bracket
14   Stay bolt

1   Caliper bolts
2   Bleed screw
3   Caliper
4   Seal
5   Piston

Fig. 8.10. Rear disc brake caliper assembly — KZ and 1978—79 US models

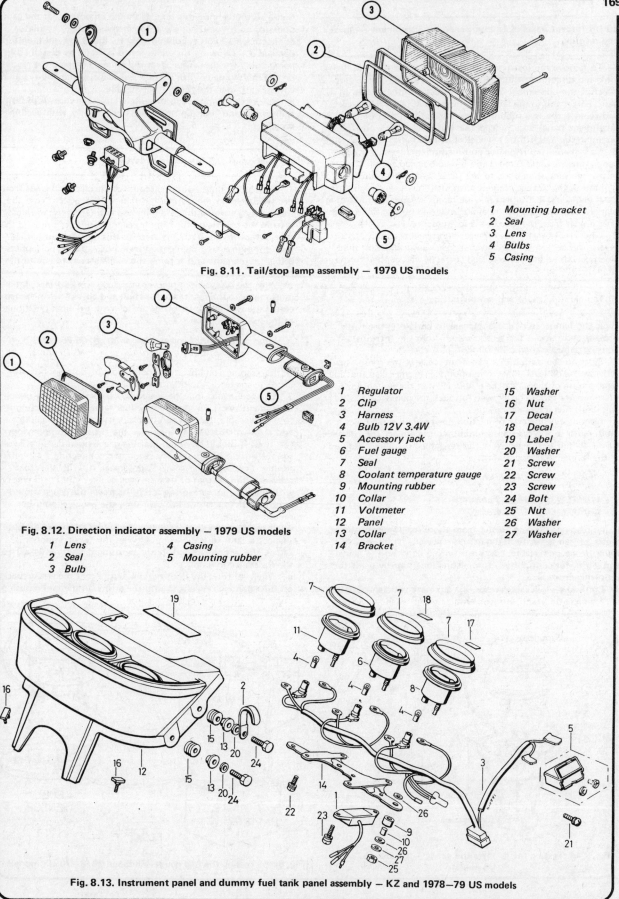

Fig. 8.11. Tail/stop lamp assembly — 1979 US models

1 Mounting bracket
2 Seal
3 Lens
4 Bulbs
5 Casing

Fig. 8.12. Direction indicator assembly — 1979 US models

1 Lens   4 Casing
2 Seal   5 Mounting rubber
3 Bulb

1	Regulator	15	Washer
2	Clip	16	Nut
3	Harness	17	Decal
4	Bulb 12V 3.4W	18	Decal
5	Accessory jack	19	Label
6	Fuel gauge	20	Washer
7	Seal	21	Screw
8	Coolant temperature gauge	22	Screw
9	Mounting rubber	23	Screw
10	Collar	24	Bolt
11	Voltmeter	25	Nut
12	Panel	26	Washer
13	Collar	27	Washer
14	Bracket		

Fig. 8.13. Instrument panel and dummy fuel tank panel assembly — KZ and 1978–79 US models

19 Instrument panel assembly: panel and instrument removal and refitting

1   To remove the instrument panel, remove the dummy fuel tank and disconnect the instrument harness block connector which is the uppermost of three such connectors located in a panel below and to the left of the instrument panel. Unscrew and remove the two mounting bolts from each side of the instrument panel base, unclip the wiring harness from the retaining clip and lift the panel clear of the frame.

2   To separate the regulator and instruments from the panel casing, unscrew and remove the four crosshead screws that secure the mounting plate to the panel casing; the plate will pull out of the casing complete with instruments and regulator. Each individual instrument may be removed from the mounting plate by removing its two retaining nuts and pulling the rubber bulb holder from the base of the instrument.

3   When refitting the instruments and panel assembly, use a reverse procedure to that given above, ensuring that the electrical wiring harness is routed correctly and clipped to the frame.

20 Temperature gauge and sensor: testing

1   If the temperature gauge appears to be faulty, drain the coolant and remove the sensor switch from the thermostat housing before testing it as follows.

2   Suspend the switch in a pan of oil so that the sensor tip is below the oil level. Place a thermometer in the oil so that the temperature of the oil can be noted. Neither the switch nor the thermometer should be allowed to touch the pan because this will result in a false reading.

3   Wearing eye and skin protection, heat the oil slowly. Set a multimeter to the resistance function and connect the probes to the switch terminals. Note the resistance readings obtained at the following temperatures:

60° C (140° F) 104.0 ohms
85° C (185° F) 43.9 ohms
110° C (232° F) 20.3 ohms
120° C (250° F) 16.1 ohms

If the readings obtained differ from those listed above, the sensor switch is faulty and must be renewed. If the switch is found to be serviceable, check the wiring between the gauge and sensor for continuity before testing the gauge by using the following procedure.

4   Connect a serviceable sensor, the auxiliary voltage regulator, a 12 volt battery and the temperature gauge as shown in the figure accompanying this text. It should be noted that the gauge operates on a 7 volt supply through the auxiliary regulator; connecting a 12 volt supply directly to the gauge will therefore damage it. Following the same test procedure as described in paragraph 2 for the sensor, heat the oil and compare the reading on the thermometer with that on the gauge. If the two readings correspond, then the gauge is serviceable.

5   Before refitting the sensor switch, clean the threads of both the switch and housing and coat them lightly with sealing compound.

21 Fuel gauge and float switch: testing

1   To test the float switch, disconnect the electrical leads from the terminals on the switch and remove the switch from the fuel tank by turning the retaining ring anti-clockwise to release it from the spigots on the tank housing. Honda recommend that special tool No HC 61076 is used for this purpose but it was found that inserting the nose ends of a pair of long-nose pliers into the ring slots and turning the pliers served to rotate the ring.

2   With the float switch placed on a clean work surface, set a multimeter to the resistance function and check the resistance across the two terminals of the switch with the float positioned as follows:

Float at bottom of travel (tank empty)   65.75 ohms
Float at top of travel (tank full)   10.15 ohms

If the readings on the multimeter do not correspond with those given above, the float switch should be renewed.

3   The fuel gauge may be tested by removing it from the instrument panel on the dummy fuel tank and connecting it to a serviceable float switch, the auxiliary voltage regulator and a 12 volt battery as shown in the figure accompanying this text. It should be noted that the gauge operates on a 7 volt supply through the auxiliary regulator; connecting a 12 volt supply directly to the gauge will damage it. With the float moved to the bottom of its travel the gauge should read empty; with the float moved to the top of its travel the gauge should read full. If this is not the case then the gauge should be renewed.

4   If both gauge and float switch are found to be serviceable, suspect a fault in the wiring between the two components. Check the wiring for continuity by using a multimeter set to the resistance function.

5   When refitting the float switch, ensure that the arrow marked on the retaining ring aligns with the arrow on the tank housing once it is fully tightened in position.

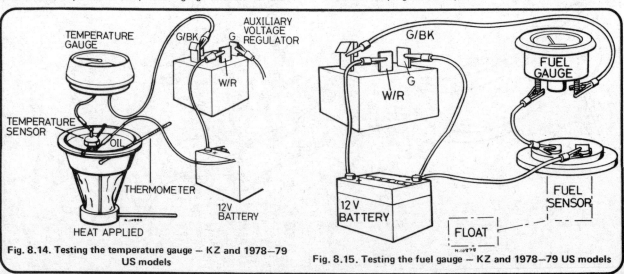

Fig. 8.14. Testing the temperature gauge — KZ and 1978–79 US models

Fig. 8.15. Testing the fuel gauge — KZ and 1978–79 US models

22 Voltmeter: testing

1   The voltmeter should be seen to read within a range of 12 to 15 volts with the engine running above 2000 rpm. If this is not so and the reading is seen to be 10 to 12 volts, suspect the battery of being excessively discharged in which case it will have to be removed from the machine and charged in accordance with the instructions given in Section 2 of Chapter 7. Alternatively, if the voltmeter is seen to be reading above 15 volts or below 10 volts, suspect a fault in the electrical system.

2   To test the voltmeter, gain access to the base of the voltmeter by detaching the instrument panel from its frame mountings. Switch on the ignition and note the reading on the scale of the voltmeter. Using a multimeter set on the voltage function on a scale of 0 − 20 (dc), place the probes of the multimeter on the terminals of the voltmeter and note the reading shown on the scale of the multimeter. If the two readings obtained are identical then the voltmeter is functioning correctly. If the readings differ then it should be assumed that the multimeter is functioning correctly in which case the voltmeter should be taken to an official Honda Service Agent or a qualified auto-electrician for testing before renewal is considered necessary.

23 Auxiliary voltage regulator: testing

1   The purpose of the auxiliary voltage regulator is to reduce the 12 volt supply provided from the battery to a 7 volt supply which is the voltage required to operate the fuel gauge and fuel sensor and the temperature gauge and temperature sensor.

2   To test the regulator, remove it from the machine as described in Section 19 of this Chapter and connect it to a 12 volt battery and a multimeter set to a voltage (dc) function as shown in the figure accompanying this text. The output of the regulator should be 7 volts; if this is not the case, renew the regulator as repair of the unit is not possible.

24 Accessory terminal box: general

1   The accessory terminal box is located beneath the left-hand side of the dummy fuel tank and is provided to power accessory equipment not exceeding a 60 Watt (5 amp) rating. The voltage supply from the terminals is 12 volt dc.

2   When connecting accessories to the terminals, ensure that the leads from the accessory equipment are connected securely to the terminals and routed clear of any engine components that will become hot or any sharp-edged cycle components upon which the leads might chafe. Retaining clips are provided on the frame either side of the terminal box and should be used where possible.

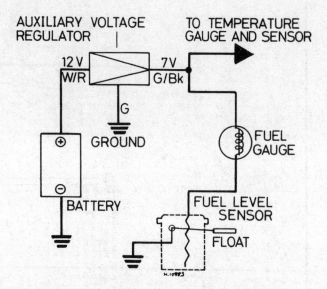

Fig. 8.16. Auxiliary voltage regulator, fuel gauge and temperature gauge circuit

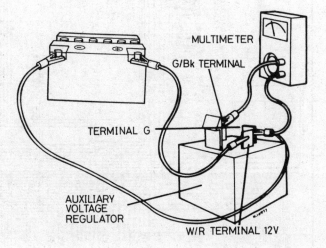

Fig. 8.17. Testing the auxiliary voltage regulator — KZ and 1978–79 US models

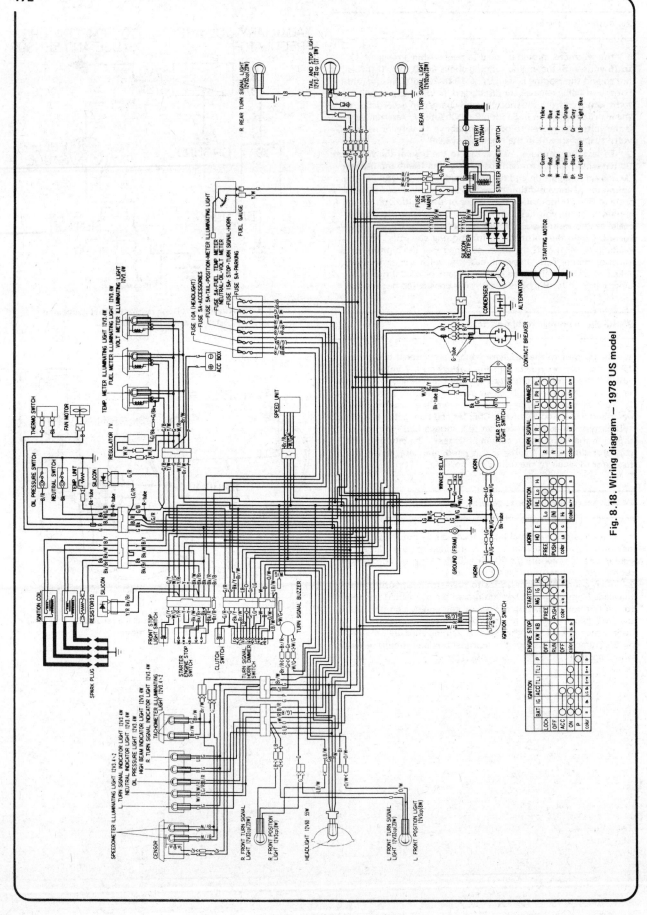

Fig. 8.18. Wiring diagram — 1978 US model

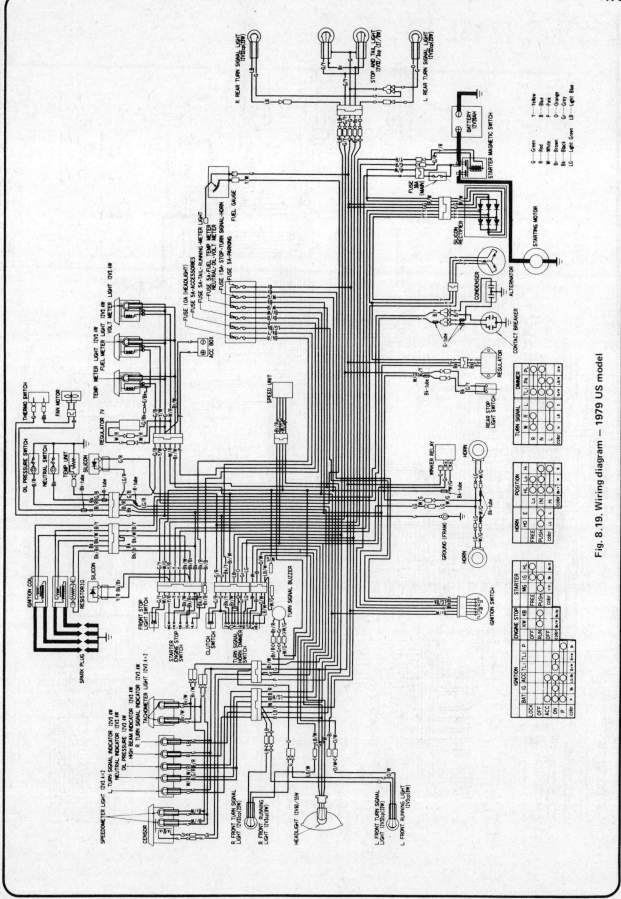

Fig. 8.19. Wiring diagram – 1979 US model

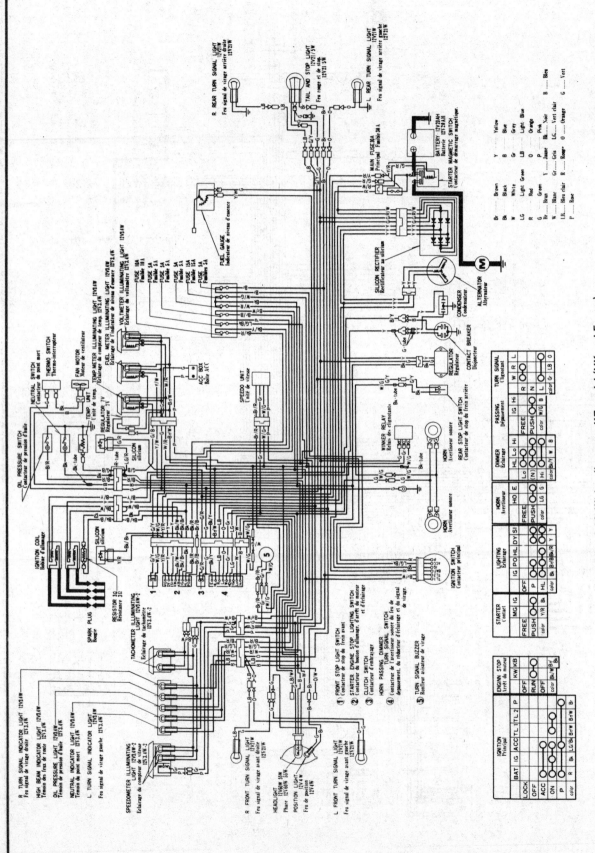

Fig. 8.20. Wiring diagram - KZ model (UK and France)

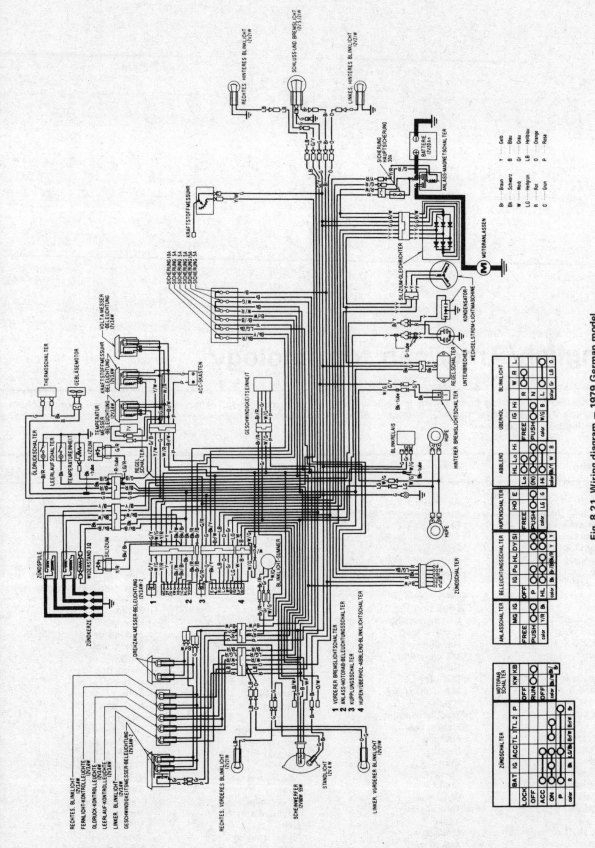

Fig. 8.21. Wiring diagram — 1979 German model

English/American terminology

Because this book has been written in England, British English component names, phrases and spellings have been used throughout. American English usage is quite often different and whereas normally no confusion should occur, a list of equivalent terminology is given below.

English	American	English	American
Air filter	Air cleaner	Number plate	License plate
Alignment (headlamp)	Aim	Output or layshaft	Countershaft
Allen screw/key	Socket screw/wrench	Panniers	Side cases
Anticlockwise	Counterclockwise	Paraffin	Kerosene
Bottom/top gear	Low/high gear	Petrol	Gasoline
Bottom/top yoke	Bottom/top triple clamp	Petrol/fuel tank	Gas tank
Bush	Bushing	Pinking	Pinging
Carburettor	Carburetor	Rear suspension unit	Rear shock absorber
Catch	Latch	Rocker cover	Valve cover
Circlip	Snap ring	Selector	Shifter
Clutch drum	Clutch housing	Self-locking pliers	Vise-grips
Dip switch	Dimmer switch	Side or parking lamp	Parking or auxiliary light
Disulphide	Disulfide	Side or prop stand	Kick stand
Dynamo	DC generator	Silencer	Muffler
Earth	Ground	Spanner	Wrench
End float	End play	Split pin	Cotter pin
Engineer's blue	Machinist's dye	Stanchion	Tube
Exhaust pipe	Header	Sulphuric	Sulfuric
Fault diagnosis	Trouble shooting	Sump	Oil pan
Float chamber	Float bowl	Swinging arm	Swingarm
Footrest	Footpeg	Tab washer	Lock washer
Fuel/petrol tap	Petcock	Top box	Trunk
Gaiter	Boot	Torch	Flashlight
Gearbox	Transmission	Two/four stroke	Two/four cycle
Gearchange	Shift	Tyre	Tire
Gudgeon pin	Wrist/piston pin	Valve collar	Valve retainer
Indicator	Turn signal	Valve collets	Valve cotters
Inlet	Intake	Vice	Vise
Input shaft or mainshaft	Mainshaft	Wheel spindle	Axle
Kickstart	Kickstarter	White spirit	Stoddard solvent
Lower leg	Slider	Windscreen	Windshield
Mudguard	Fender		

Conversion factors

Length (distance)
Inches (in)	X 25.4 = Millimetres (mm)	X 0.039 = Inches (in)
Feet (ft)	X 0.305 = Metres (m)	X 3.281 = Feet (ft)
Miles	X 1.609 = Kilometres (km)	X 0.621 = Miles

Volume (capacity)
Cubic inches (cu in; in^3)	X 16.387 = Cubic centimetres (cc; cm^3)	X 0.061 = Cubic inches (cu in; in^3)
Imperial pints (Imp pt)	X 0.568 = Litres (l)	X 1.76 = Imperial pints (Imp pt)
Imperial quarts (Imp qt)	X 1.137 = Litres (l)	X 0.88 = Imperial quarts (Imp qt)
Imperial quarts (Imp qt)	X 1.201 = US quarts (US qt)	X 0.833 = Imperial quarts (Imp qt)
US quarts (US qt)	X 0.946 = Litres (l)	X 1.057 = US quarts (US qt)
Imperial gallons (Imp gal)	X 4.546 = Litres (l)	X 0.22 = Imperial gallons (Imp gal)
Imperial gallons (Imp gal)	X 1.201 = US gallons (US gal)	X 0.833 = Imperial gallons (Imp gal)
US gallons (US gal)	X 3.785 = Litres (l)	X 0.264 = US gallons (US gal)

Mass (weight)
Ounces (oz)	X 28.35 = Grams (g)	X 0.035 = Ounces (oz)
Pounds (lb)	X 0.454 = Kilograms (kg)	X 2.205 = Pounds (lb)

Force
Ounces-force (ozf; oz)	X 0.278 = Newtons (N)	X 3.6 = Ounces-force (ozf; oz)
Pounds-force (lbf; lb)	X 4.448 = Newtons (N)	X 0.225 = Pounds-force (lbf; lb)
Newtons (N)	X 0.1 = Kilograms-force (kgf; kg)	X 9.81 = Newtons (N)

Pressure
Pounds-force per square inch (psi; lbf/in^2; lb/in^2)	X 0.070 = Kilograms-force per square centimetre (kgf/cm^2; kg/cm^2)	X 14.223 = Pounds-force per square inch (psi; lbf/in^2; lb/in^2)
Pounds-force per square inch (psi; lbf/in^2; lb/in^2)	X 0.068 = Atmospheres (atm)	X 14.696 = Pounds-force per square inch (psi; lbf/in^2; lb/in^2)
Pounds-force per square inch (psi; lbf/in^2; lb/in^2)	X 0.069 = Bars	X 14.5 = Pounds-force per square inch (psi; lbf/in^2; lb/in^2)
Pounds-force per square inch (psi; lbf/in^2; lb/in^2)	X 6.895 = Kilopascals (kPa)	X 0.145 = Pounds-force per square inch (psi; lbf/in^2; lb/in^2)
Kilopascals (kPa)	X 0.01 = Kilograms-force per square centimetre (kgf/cm^2; kg/cm^2)	X 98.1 = Kilopascals (kPa)

Torque (moment of force)
Pounds-force inches (lbf in; lb in)	X 1.152 = Kilograms-force centimetre (kgf cm; kg cm)	X 0.868 = Pounds-force inches (lbf in; lb in)
Pounds-force inches (lbf in; lb in)	X 0.113 = Newton metres (Nm)	X 8.85 = Pounds-force inches (lbf in; lb in)
Pounds-force inches (lbf in; lb in)	X 0.083 = Pounds-force feet (lbf ft; lb ft)	X 12 = Pounds-force inches (lbf in; lb in)
Pounds-force feet (lbf ft; lb ft)	X 0.138 = Kilograms-force metres (kgf m; kg m)	X 7.233 = Pounds-force feet (lbf ft; lb ft)
Pounds-force feet (lbf ft; lb ft)	X 1.356 = Newton metres (Nm)	X 0.738 = Pounds-force feet (lbf ft; lb ft)
Newton metres (Nm)	X 0.102 = Kilograms-force metres (kgf m; kg m)	X 9.804 = Newton metres (Nm)

Power
Horsepower (hp)	X 745.7 = Watts (W)	X 0.0013 = Horsepower (hp)

Velocity (speed)
Miles per hour (miles/hr; mph)	X 1.609 = Kilometres per hour (km/hr; kph)	X 0.621 = Miles per hour (miles/hr; mph)

Fuel consumption*
Miles per gallon, Imperial (mpg)	X 0.354 = Kilometres per litre (km/l)	X 2.825 = Miles per gallon, Imperial (mpg)
Miles per gallon, US (mpg)	X 0.425 = Kilometres per litre (km/l)	X 2.352 = Miles per gallon, US (mpg)

Temperature
Degrees Fahrenheit (°F) $= (°C \times \frac{9}{5}) + 32$

Degrees Celsius (Degrees Centigrade; °C) $= (°F - 32) \times \frac{5}{9}$

*It is common practice to convert from miles per gallon (mpg) to litres/100 kilometres (l/100km), where mpg (Imperial) x l/100 km = 282 and mpg (US) x l/100 km = 235

Index